Physiology and Pathology of Membrane Digestion

By Aleksandr M. Ugolev

Director, Laboratory of Physiology of Nutrition
I. P. Pavlov Institute of Physiology
Academy of Sciences of the USSR, Moscow

Translated from Russian by

Jacob A. Stekol

Head, Department of Physiological Chemistry and Nutrition
The Institute for Cancer Research
Fox Chase, Philadelphia, Pennsylvania

Springer Science+Business Media, LLC 1968

Aleksandr Mikhailovich Ugolev was born in 1926 in Leningrad and is a graduate of the Naval-Military Medical Academy and of the Dnepropetrovsk Medical Institute. Having worked in the laboratory of the well-known biochemist A. E. Braunshtein and having taught for five years at the Leningrad Medical Institute of Stomatology, in 1955 he became the Scientific Collaborator of the Laboratory of General Physiology at the Institute of Normal and Pathological Physiology of the Academy of Medical Sciences. In 1958, Ugolev earned the degree of Doctor of Science and since 1963 has been Director of the Laboratory of Physiology of the I. P. Pavlov Institute of Physiology. A recipient of the coveted Pavlov Award for his work in the field of the physiology of digestion, Ugolev is a corresponding member of the Academy of Sciences of the USSR.

The original Russian text was published by the Leningrad Division of Nauka Press in 1967 for the Joint Scientific Council on Human and Animal Physiology of the Academy of Sciences of the USSR and has been corrected by the author for this edition.

Александр Михайлович Уголев

ФИЗИОЛОГИЯ И ПАТОЛОГИЯ ПРИСТЕНОЧНОГО (КОНТАКТНОГО) ПИЩЕВАРЕНИЯ

FIZIOLOGIYA I PATOLOGIYA PRISTENOCHNOGO (KONTAKTNOGO) PISHCHEVARENIYA

Library of Congress Catalog Card Number 68-19186

ISBN 978-1-4899-6202-7 ISBN 978-1-4899-6409-0 (eBook)
DOI 10.1007/978-1-4899-6409-0

Originally published by Plenum Press in 1968.
Softcover reprint of the hardcover 1st edition 1968

Physiology and Pathology of Membrane Digestion

Foreword

The physiology and pathology of membrane (contact or surface) digestion of the three main components of the food of animals and man is a rapidly developing field, and there is little doubt that newer observations will become available as time progresses. As the discoverer of the phenomenon of membrane digestion, Professor Ugolev is well aware of the relative infancy of this subject, and in presenting the book at this time his primary purpose was to acquaint the reader with the experimental basis for the observed phenomenon and to stimulate interest in the further work required for the elucidation of details and for proper orientation.

Translations as means of communication of ideas and experimental facts are of obvious importance and they require no apologies or explanations, particularly to those who are engaged in scientific endeavors. In making this book available to those who find the Russian language an insurmountable barrier, our main efforts were directed toward preserving the author's meaning, spirit, and intent as accurately as possible. It is hoped that this objective has been attained, and that the minor unintentional errors will be forgiven by the reader and by the author.

The translator wishes to express to Professor Ugolev, whom he met in Leningrad in 1965, his appreciation for the trust in the ability of the translator to make Professor Ugolev's labor of love, the present book, available in the English language.

March 1968 Jakob A. Stekol

Preface to the English Edition

I gratefully accepted the offer of Plenum Press to prepare a special preface for the English translation of this book. However, as I write it, I am experiencing a feeling, familiar to all investigators, that everything I have done and written could have been done and written a little better.

In 1958 a series of facts forced me to come to two conclusions which until then I had carefully avoided: (1) There are three, and not two, types of digestion; in addition to the previously known intracellular and extracellular digestion, there also is membrane (contact) digestion. (2) The classic scheme of the activity of the digestive system in higher animals (cavital digestion–absorption) is insufficient. In reality the transformation of foodstuffs occurs by the scheme: cavital digestion–membrane digestion–absorption.

In short, there exists a process of digestion which takes place on the surface of cell membranes (in man and in higher animals these are the membranes of the microvilli of the intestinal epithelium). Thus, on the border of the extracellular and intracellular media there is not only an active transport of substances but also their depolymerization.

As will be shown in detail in this book, the hydrolysis of foodstuffs at the time of their contact with the enzymes localized in the membrane (hence the name "membrane" or "contact" digestion) differs materially from the other types of digestion in its mechanisms and possesses several biological peculiarities, including the possibility of close coordination of digestive and transport processes.

Regardless of the real role of membrane digestion, we had to investigate it in order to understand the evolution of the mechanisms which ensure the transformation of foodstuffs. But when it became clear that digestion in man and higher animals was mainly membrane digestion and that what had been considered the whole truth was only part of the truth, it did not take much imagination to come to the conclusion that these were only the first steps in the development of a field which is essentially limitless.

The first publications of this new concept in Russian and English were not met by widespread enthusiasm. The resulting solitude, however, was not unendurable, thanks to the encouragement and support of such scientists as Academicians A.E. Braunshtein and V.N. Chernigovskii. Their criticism and advice, and the certainty that further work in this field was my duty, were extremely important in the further development of our investigations. Three years later, membrane digestion was already under study by a group of collaborators. The reader will constantly encounter their names in this book.

As soon as the fact of the existence of membrane digestion ceased to elicit doubt, at least in those who studied it experimentally, we made an attempt to prepare our first book wholly devoted to membrane digestion. We had to consider various aspects of the problem, and these were many: physiology of membrane digestion; physics and chemistry of membrane digestion; enzymology of membrane digestion; ontogenesis and phylogenesis of membrane digestion; interrelationship of membrane and cavital digestion; interrelationship of membrane digestion and absorption; membrane digestion and pathology of the gastrointestinal tract; etc.

Any one or several of these problems could have been the subject of a separate review. I was convinced, however, that it would be more useful to consider all these problems together, even at the risk of being brief and furnishing an inadequate analysis of each.

We made three such attempts: the first monograph in Russian appeared in 1963; in 1965 we published a review in Physiological Reviews; and in 1967 we published the present monograph.

In reviewing the often widely separate aspects of membrane digestion, I gratefully recalled the names of my teachers, Academician V.N. Chernigovskii, who taught me general physiology and neurophysiology, Professor A.D. Slonim (ecological and comparative physiology), Academician A.E. Braunshtein (enzymology), and Professor N.N. Samarin (gastroenterology and experimental surgery). It goes without saying that I do not mention their names in order to conceal the weaknesses of this book under the mantle of their scientific accomplishments.

Finally, I hope that the author's shortcomings will not prevent the reader from seeing the importance of membrane digestion and its significance to biology and medicine.

* * *

In our time no work is the result of efforts of one person. In the Introduction to the original edition I gratefully mention several names.

Here it gives me great pleasure to express thanks to all my colleagues without whose investigations the concept of membrane digestion would have been impossible. I am deeply touched that Dr. Stekol took upon himself the difficult and thankless job of translating this book into English.

A. M. Ugolev

Preface

This volume has evolved from an earlier one, published four years ago, in which the first attempt was made to systematically analyze the basic problems related to the physiology and pathology of membrane digestion. If the time elapsed seems rather brief, this can be attributed to the rapid and significant development of the subject.

It is worthwhile to keep in mind that the development of any problem is accompanied not only by new facts and hypotheses but also by changes in the methods of evaluation of the results obtained. During the past few years analytical criteria have become considerably more severe and exact, and we note with some satisfaction that in spite of this, detailed analysis of available data has provided new and strong arguments supporting the important role of membrane digestion and also has permitted resolution of many previously debatable points.

In the present volume the various questions related to membrane digestion are treated from the standpoint of a person who regards this mechanism not only indifferently but skeptically or even somewhat antagonistically, but not with such fanaticism as to fail to see the facts and lose the capacity to analyze them. My colleagues noted this while reading the manuscript and suggested that this style may leave a certain sense of uncertainty in the reader's mind, but perhaps the reader will be compensated for this by the realization that the existence and the importance of membrane digestion are independent of the enthusiasm of the author.

The chapters of this book deal with the various rapidly developing aspects of the physiology and pathology of membrane digestion, each of which could be the subject of a separate monograph. The brevity of the book is thus due to lack of space and not to scarcity of available data.

We shall now give a brief description of the contents of the book. In Chapter I, an attempt is made to define the basic types of digestion and, what is less important, to explain why they have been preserved and developed in the evolutionary process of the animal world and why an organism often employs two or even three types of digestion, not limiting itself to one type, the most "perfect" one.

Chapter II deals with intestinal digestion in mammals and the role of membrane hydrolysis in the performance of this important stage of treatment and assimilation of foodstuffs. In the characterization of views widely held in contemporary scientific literature, we, like others, were confronted by the absence of a thorough evaluation of experimental data. It appeared that many contradictions could be removed if a more exact description of the methods employed by various authors was obtained. Many of the contradictions thus proved to be apparent rather than real. This made it possible in the concluding section of Chapter II not only to appreciate the latest findings which define the role of membrane digestion in the activity of the gastrointestinal tract of mammals, but to present a picture which integrates all the known facts in this field.

Chapter III contains a very brief account of available information concerning the physicochemical and structural aspects of membrane digestion. We wanted to make this chapter broader, but since several substantial monographs have appeared which are devoted to membranes and the physical chemistry of biological surfaces, we limited ourselves to important information which is necessary to the understanding of membrane digestion and which is absent in these monographs. In addition, we included in this chapter more recent data obtained in our laboratory as well as hypotheses which may, perhaps, be useful later.

Chapter IV is devoted to a detailed description of membrane digestion, and Chapter V to its clinical aspects. Evolutionary problems of membrane digestion (Chapter VI) are treated relatively briefly in this book, but the data cited not only are important for the appreciation of the wide distribution of this mechanism among living organisms at various stages of the evolutionary ladder, but may prove useful to those who are interested in the comparative physiology of membrane digestion.

Although the existence of membrane digestion is no longer questioned, the author considered it his duty to analyze not only the arguments which prove the existence of this mechanism but also the arguments (hypotheses for the most part) developed by the opponents. Throughout, we have attempted to differentiate between facts and the conclusions derived from them via hypotheses (which are more useful the more one distrusts them).

In this book the terms "wall," "contact," or "membrane" digestion are used synonymously. All three were proposed by the author in 1958, but the last one apparently can best be justified from a cytological point of view.

It is a pleasure to express my gratitude to a wide circle of investigators. First, my co-workers and friends: M.I. Dumesh, N.N. Iezuitova, R.I. Kushak, T. Ya. Nadirova, R.A. Ovdeichuk, K.R. Rakhimov, N.M. Timofeeva, M.Yu. Chernyakhovskaya, and K.I. Khalimov, as well as V.V. Chernousova (Bashkir State University), A.T. Stepanova (Pyatigorsk Pharmaceutical Institute), N.V. Toropova and Yu.M. Toropov (Frunze), and I.K. Salenietse, M.K. Marausk, Ya. Ya. Nursk, V.G. Kirs, G.G. Lapteva, and E.M. Ustinkova (Latvian State University). I am also grateful to Dr. P. De Laey (Belgium), with whom I collaborated for many years and to whom belongs the credit for significant contributions to the development of membrane digestion, and to Dr. O. Koldovsky and his associates (Czechoslovakia). Collaboration of our laboratories was stimulating to the solution of many important problems.

I am deeply indebted to N.N. Iezuitova, V.G. Kassil, G.V. Makukhina, N.M. Timofeeva, E.N. Yakobson, and particularly to my wife, S.V. Ugoleva, for aid in the preparation of this book.

Valuable comments during the preparation of this book were made by Professors G.P. Conradi and V.I. Rosengardt, and Dr. M.N. Maslova.

The first critics of our initial investigations of membrane digestion were Academicians V.N. Chernigovskii and A.E. Braunshtein, to whom I am grateful for the valuable advice which was so crucial to all our subsequent work.

Finally, I am grateful always to my teachers, Academician V.N. Chernigovskii and Professor A.D. Slonim.

Contents

Chapter I

Basic Types of Digestion and Their Classification

In this preliminary outline no attempt is made to give a complete description of all the digestive processes. We are interested in a more immediate problem—defining concepts such as cavital, intracellular, or membrane digestion and analyzing the possible role of each of these basic types of digestion.

The concept "digestion" is employed in its classical meaning, by the sum total of processes which ensure enzymatic cleavage (mainly hydrolytic) of polymeric nutrients to their basic components, basically monomers (Mali, 1886; Pavlov, 1897; Starling, 1906; London, 1916; Bykov, 1940; Koshtoyants, 1950; Buddenbrock, 1956; Shtraub, 1963; Bockus, 1964; and others).

Digestion is a necessary stage in the utilization of organic polymers, and since the majority of contemporary heterothrophic organisms are nourished by more or less complex organic substances, these organisms must possess the developed systems which perform the digestion.

Digestion is invariably terminated by the enzymatic cleavage of the polymers, but the mechanism of this process varies, and one can distinguish several basic types of digestion. It appears one must distinguish three types of digestion, depending on the source of the enzymes which accomplish the process. Most frequently the enzymes are produced by the organism which assimilates the food. This, for example, is true for higher animals. This type can appropriately be called inborn digestion. However, even in higher animals (such as the ruminants and many rodents), symbiotic organisms (bacteria and protozoa) which produce a series of enzymes play an important role in digestion. This type of digestion is particularly important in cases in which the nutrients are not cleaved by the enzymes of the host and was named symbiotic digestion. It is widespread also among the lower animals, for example, *Arthropoda*. A detailed review of this problem is available (Buddenbrock, 1956; Ugolev, 1961a). The enzymes which these organisms contain play a definite role in the destruction of nutritive polymers. This type of digestion is called autolytic. It is possible that autolytic digestion was the earliest in the history of the evolution of the animal world, and it has not completely lost its importance in the nutrition of contemporary animals. It is possible, for instance, that in the utilization of fresh foodstuffs by the *Herbivora*, digestion, in part, is accomplished by the enzymes present in plants. Curing of meat is nothing else but partial autolysis which, from its significance, can be considered as partial digestion.

Active or inborn digestion underwent the greatest evolution and distribution in the animal world. Even where another type of digestion prevails (as symbiotic digestion in ruminants), the inborn digestion is the final one and is an extremely important stage in the utilization of foodstuffs.

Thus far we have discussed the origins of digestive enzymes. Another classification (the most widespread) is based on the localization of the digestive processes.

Comparatively recently two basic types of digestion were recognized. Thus, in 1950 Koshtoyants wrote: A review of the basic forms of digestion of modern animals distinguishes intracellular and extracellular digestion. From a comparative-morphologic standpoint one can distinguish extracellular (intestinal) digestion, which takes place in a specially differentiated system of the digestive tract and is characteristic of higher multicellular organisms, and the digestive process (intracellular) which occurs in the protoplasm of the cell. The latter type, in multicellular ones which are phylogenetically more ancient and either do not possess the digestive tube or in which the tube is not sufficiently differentiated. To this group belong the following multicellular organisms: sponges, *Coelenterata*, certain flat worms, and, among the higher multicellular organisms, certain mollusks.

In mollusks the processes of intracellular digestion coexist with extracellular digestion. The intracellular digestion is absent in worms, *Coelenterata*, and vertebrate animals (at least in the adult species of these animals). The basic difference between these two types of digestion consists in the fact that in the case of intracellular digestion the enzymes are acting on the nutrients inside the cells, whereas in the case of extracellular digestion these enzymes formed by the protoplasm of special cellular glands are secreted into the intestinal canal where the cleavage of the nutrients takes place. Similar views were expressed by Barrington (1962) in his review.

Ten to twenty years ago such a classification appeared to be not only satisfactory but complete as well. For this reason the concepts of intracellular and contact digestion or extracellular and distal digestion appeared to be synonymous.

In the past few decades our concept of the cell as a microheterogeneous system progressed rapidly. One of the most fruitful results of this progress was a better understanding of the role of cellular membranes in various processes, particularly those in which a whole assembly of enzymes and not individual ones participate. It became evident that various enzymatic systems exist in hyoplasm, not in solution or in the matrix of individual organoids, but included in the composition of cell membranes as systems organized in space. Reviews related to this problem can be found in multivolumed texts—"The Cell" by Brachet and Mirsky in volume II (1961) and in "Manual in Cytology" (1965). The foregoing relates not only to mitochondria but also to the cell membrane which divides the intracellular and extracellular media. Thus, one can see that the traditional division of digestive processes is not complete in that it does not consider that fine border which separates the cell from the surrounding medium. The cell membrane, as has been discovered by modern research technology, turned out to be quite a complex structure endowed with various enzymatic activities (see reviews by Rothstein, 1954; Booij, 1962; Komissarchik, 1965; Rumyantsev, 1965; Ugolev, 1965b; Crane, 1966; and others). It is at this point that we encounter contact (or membrane, as they began calling it in the West) digestion, the analysis of which is the subject matter of this book.

From the foregoing it is clear that membrane digestion is an intermediate type of digestion. Indeed, insofar as it takes place on the cell membrane at the moment the nutrient substrates contact the cell surface, this type of digestion can be called contact digestion. One will recall that earlier contact digestion was called intracellular digestion. Since membrane digestion occurs before nutrient substrates penetrate the cell, it is in reality extracellular and from this standpoint it resembles the classical extracellular digestion.

Membrane digestion is induced by enzymes which are part of the cell structure, and in this sense it can produce effects which resemble those of intracellular digestion. At the same time it possesses a series of specific characteristics.

These problems are discussed in detail in subsequent chapters of this book, but it is evident even at this point that there is a serious basis for the existence of a third basic type of digestion.

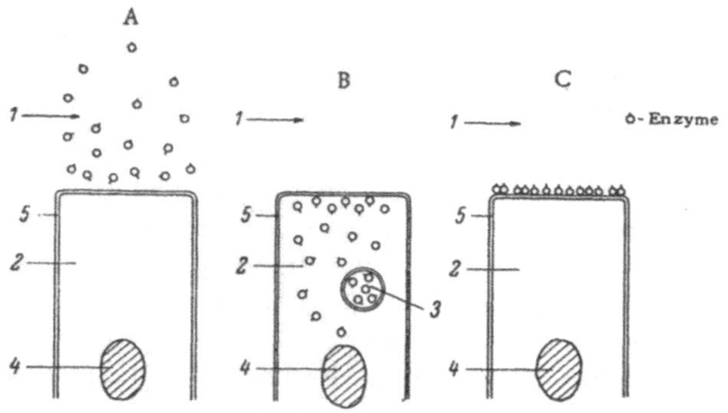

Fig. 1. Localization of hydrolysis of food during various types of digestion. (A) Extracellular; (B) intracellular; (C) membrane (contact). (1) Extracellular fluid; (2) intracellular fluid; (3) intracellular vacuole; (4) nucleus; (5) cell membrane.

We shall now attempt a brief description of the basic types of digestion (Fig. 1).

1. **Extracellular Distal Digestion** (see reviews: Pavlov, 1897; London, 1916; Babkin, 1927; 1960; Buddenbrock, 1956; Bockus, 1964). Since outside the cell two types of digestion proceed, this designation seems appropriate in order not to confuse digestion on the surface of cell membranes with digestion in the medium which surrounds the cell. Extracellular distal digestion is characterized by the fact that the enzymes which are synthesized in the cell are transported to the extracellular medium and they act at some distance away from the secreting cells. In the majority of highly developed animals, the secreting cells (for instance, pancreatic and salivary glands of mammals) are located some distance away from where the digestive enzymes act.

Whenever distal digestion takes place in a special region or cavity, it is generally spoken of as cavital digestion. But it is well known (Kuznetsov, 1948; Koshtoyants, 1950; Buddenbrock, 1956; and others) that in many instances distal digestion is performed outside the organism which produced the enzymes. Thus, many insects introduce digestive enzymes into the immobilized victim, and in these cases distal digestion is not a cavital digestion. Bacteria secrete various digestive enzymes (Stefenson, 1951; Ugolev, 1961a) into the culture medium.

If one considers distal digestion at the molecular level, one observes the following picture. The enzymes are dissolved in liquid phase; their transfer is determined by the laws of heat movement (of considerable importance could be the adsorption of food substrates).

In cases where several enzymes act in solution, their distribution depends on probability. Since the enzymes are in solution, any orientation of their active centers is possible. It is very important to keep the latter in mind in comparing the properties of distal (extracellular) and membrane (contact) types of digestion.

2. **Intracellular Digestion** (Jordan, 1934; Koshtoyants, 1950; Buddenbrock, 1956; Müller et al., 1963). By this term are designated all cases in which intact or partially cleaved food substrate penetrates the cell and there undergoes further hydrolysis. This type of digestion is widely distributed among monocellular and lower organisms, and this fact provided the grounds to consider this type of digestion the most primitive and oldest from the evolutionary standpoint (see reviews: Jordan, 1934; Koshtoyants, 1950). However, it is now known that the intracellular digestion mechanism is widely distributed. We speak not only of phagocytic properties of leucocytes and reticuloendothelial system, but also of the fact that one of the varieties of phagocytosis (pinocytosis) is more or less a peculiarity of cells of higher animals, and is of ectodermal and endodermal origin. Intracellular digestion can be performed in special intracellular compartments (digestive vacuoles) which are always present or which can be formed during phagocytosis and pinocytosis (see reviews: Novikoff, 1961; Kheisin, 1965), and disappear after the completion of cleavage of the foodstuff.

De Duve (1959), and many authors later (see reviews: Novikoff, 1961; Mashanskii, 1965), observed special organoids which contained a considerable conglomeration of hydrolytic enzymes, lysosomes. It is assumed that lysosomes can participate in intracellular digestion by coming in close contact with and entering the pinocytotic vacuole where the foodstuff substrates are dissolved. It is clear that under these circumstances intracellular extraplasmic digestion takes place. It should be noted that the majority of cases which were previously considered as digestion in the cytoplasm, turned out, under electron microscope examination, to be extracytoplasmic digestion. Electron microscopy revealed the presence of very small pinocytotic vacuoles surrounded by a cell membrane.

Thus, intracellular digestion is always or almost always extraplasmic and on closer examination remarkably resembles distal digestion since in both cases the enzymes act in the extracellular medium. Also around the vacuole there is a typical cell barrier and the physicochemical regularities of this process differ very little from digestion which takes place in macroregions. However, biologically these processes are substantially different.

3. **Membrane Digestion.** This process is performed by enzymes which are fixed on the cell membrane, and it occurs on the border between the extracellular and intracellular medium. In the physicochemical aspect, membrane digestion resembles heterogeneous catalysis on dissimilar surfaces. The enzymes are fixed; their active centers are oriented in a definite manner toward the surface of the membrane and water phase. For this reason, free orientation of active centers toward foodstuff substrates is impossible. Deeply seated bonds apparently are inaccessible to the action of enzymes which effect membrane digestion. By this property membrane digestion materially differs from cavital and intracellular digestion whenever the latter takes place under the influence of enzymes which are dissolved in vacuoles. In the case of membrane digestion, organization in space of enzyme ensembles is possible. Finally, effective interaction between the processes of hydrolysis and transmembranal shift takes place on the cell membrane surface. Thus (and this is shown in detail later), the cell membrane may serve not only as a transport surface but also as a digestive surface. It would be appropriate to mention here that frequently (particularly in foreign literature) the concept of digestive surface is interpreted rather arbitrarily. As a first approximation, the entire intestinal epithelium could be considered as the inner surface of the small intestine. Then intracellular hydrolysis does not exclude the concept that the intestinal surface is of digestive nature. Certainly, however, such use of terms is rather free, to say the least.

Membrane digestion, as a third type of digestion, is characterized not by the proximity of enzymatic hydrolysis to the surface of cell membranes but by the fact that these processes occur on the membranes which divide two environments, intracellular and extracellular, as well as by the fact that the enzymes which effect the membrane digestion are fixed on the cell membranes.

Later on we describe in detail the structures which effect membrane digestion. For the present we wish to note that in certain cases these are highly specialized cell surfaces (for instance, the brush border of the small intestine of higher animals), while in others it is one of the simplest cell membranes, as it exists in yeasts.

We have briefly described the basic types of digestion. It is important to emphasize, however, that among these there exist many transitions. As an example, vacuole digestion, which we consider traditionally as one of the varieties of intracellular digestion, in many cases can be considered as a miniature variant of distal digestion taking place within the cell. This is legitimate in cases in which enzymes are secreted into the vacuole and effect their action in solution.

It is also generally considered that during pinocytosis a portion of the cell outer membrane is invaginated inside the cell along with the enzymes fixed on it. If hydrolysis takes place on the inner surface of the pinocytotic vacuole, then it could be assumed that membrane hydrolysis, shifted inwardly, takes place. Another example is the case of alkaline phosphatase which, as has been shown by Novikoff (1961) and others, is fixed on the inner surface of lysosomes.

Having considered the three basic types of digestion, we must now return to the question which was posed first in the 19th century and which, as can be seen from the quotation from Koshtoyants, retained its reality up to the present time.

Which of the types of digestion should be considered the most primitive? Which of these turned out to be the most perfect and assumed the dominant position in higher animals? We discussed these questions in detail elsewhere (Ugolev, 1961a, 1963). It can be said with certainty at present that all three types of mechanisms of hydrolysis of foodstuffs play a role at all stages of the evolutionary ladder. Moreover, it became apparent that in some closely related species the hydrolysis of the same substrate, sucrose, for example, can be accomplished by distal as well as membrane digestion. It is probable that each type of basic digestion possesses its own peculiar advantages which permitted not only its preservation but progress in evolutionary development as well. But each of these types had its limitations; this made it impossible to substitute one type of digestion for another in all animals consuming all types of nourishment. Moreover, the advantages as well as the limitations peculiar to distal, membrane, and intracellular types of digestion led to the situation that during the evolutionary process the majority of organisms began using them jointly. As is shown later, frequently one can observe that the organism employs two or even three different types of digestion. In Chapter VI we consider in detail the interrelationship between cavital and membrane digestion in higher animals and show that because of the interrelationship of these mechanisms a high degree of effectiveness and economy of work by the gastrointestinal tract is accomplished. It may be hoped that in time, when in the fields of bionics and engineering biology it will become possible to describe exactly the structure of living systems, it will also become possible to formulate the potentialities and limitations of the distal, membrane, and intracellular types of digestion.

But even today it is possible to give an approximate description of this problem. During distal digestion there is no space limitation. The enzymes which are secreted into the extracellular fluid can act on large food particles which cannot be introduced inside the cell or digested on its surface. A good example of the unique advantages of distal digestion is the digestion of plant cells and of organisms which are covered with protective coverings. The entry of food objects inside the cell can be considered as deposition of foodstuffs by the cell. In this sense intracellular digestion cannot be reproduced by either cavital or membrane digestion.

Membrane digestion ensures a very effective interaction between the hydrolytic and transport processes. No less important is the circumstance that membrane digestion, by virtue of the possession of advantages of intracellular digestion, occurs directly on the cell. Because of this the probability that products of hydrolysis are assimilated is greater in the case of membrane digestion than in the case of cavital digestion. Certainly, of great importance is the circumstance that the surface enzymes are organized in space. However, membrane digestion becomes poorly effective when hydrolyzable bonds are covered inside the molecules of foodstuffs and are inaccessible to the action of the enzymes fixed on the surface.

The aforesaid does not completely cover the specific features of the three types of digestion, but it is sufficient for the assumption that in the gastrointestinal tract of higher animals the initial stages of hydrolysis will be taking place at the expense of cavital digestion, and the final ones, by membrane digestion. As will be shown later, indeed in the stomach of omnivorous animals membrane digestion is absent and only cavital digestion exists, while in the intestine both mechanisms effectively interact.

The basic types of digestion which exist in heterotrophic organisms are sufficiently differentiated, and the investigator is constantly confronted with the problem of which one he is dealing with. We know too little as yet to predict with certainty which type of digestion may predominate in animals which belong to this or that group with this or that nature of nourishment. For this reason it is always necessary to conduct special experiments to answer this question. Occasionally the problem is easily solved, and often with great difficulty.

Most often one is confronted by combinations of various types of digestion in the same organism and, probably, in this is concealed one of the secrets of the effectiveness and economy of the digestive apparatus. About thirty years ago Barcroft (1937) and Ukhtomskii (1941), while defining integration, discussed the superiority of integrative activity which consists in the fact that several parameters, which individually produce small effect, collectively produce much greater effect than their arithmetic sum. This is particularly true when non-equivalent parameters are integrated and the system of these parameters loses all the drawbacks of each, thus acquiring and increasing their advantages. This is fully relevant to membrane digestion and its interaction with other mechanisms which ensure the transformation and utilization of foodstuffs.

Chapter II

Digestive Processes in the Small Intestines of Higher Animals. The Role of Membrane Digestion

The basic processes of cleavage of food polymers and the absorption of organic and inorganic food components, a considerable amount of water, salts, and certain organic materials which enter together with digestive juices take place in the small intestine.

As is known, in the oral cavity mostly mechanical breakdown of food and its hydration with saliva take place. Acid denaturation of proteins (under the influence of hydrochloric acid of stomach secretions) and initial stages of peptic digestion of foodstuffs take place in the stomach. Thus, almost all the hydrolysis of proteins and all the hydrolysis of fats, polysaccharides, and nucleic acids occur in the small intestines.

Since pancreatic juice and bile as well as the secretions of Brunner glands enter the intestines, it was initially assumed that these perform the hydrolysis of foodstuffs in the intestinal cavity. It does not seem necessary to describe in detail the classical picture of digestion in the small intestine. This has been done in our previous book (Ugolev, 1963) and in monographs of Palmer (1963), Bockus (1964), and in a recent summary by Täufel (1965). For this reason in this chapter we present data pertaining to the involvement of the intestinal surface in the enzymatic transformations of foodstuffs.

First we will examine the available facts and assumptions regarding the involvement of enzymes which are located on the surface of intestinal cells in the cleavage of various foodstuffs. Then we will analyze certain contradictions which arose in this field. Finally, we will present a series of the latest data which, as it appears to us, may give a sufficiently adequate answer regarding the existence and the role of membrane digestion in the processes which take place in the small intestine of higher animals.

HYDROLYSIS OF POLYSACCHARIDES

The majority of investigators who studied the hydrolysis of starch in the small intestine of animals and man tend to agree that the degradation of starch to oligosaccharides is induced by the enzymes of pancreatic juice acting in the small intestine (Borgström et al., 1957; Borgström and Dahlqvist, 1958; Dahlqvist and Borgström, 1961; and others).

However, in 1960-61 we found that the rate of enzymatic hydrolysis of starch during the perfusion of intestinal slices of white rats or during the incubation of inverted sections* is several times smaller than the rate of hydrolysis of starch at the moment of its passage across the intestine.

Since the conditions of the medium during perfusion and during the determination of enzymatic activity in vitro were identical, we concluded that the hydrolysis in the intestine is effected not only by the enzymes present in the intestinal contents but also by the enzymes which are bound to the structure of the small intestine. Here, as in many other cases, it could be assumed that the hydrolysis of starch takes place either on the outer surface of the small intestine or intracellularly.

The second assumption, however, is not realistic since not only starch but also dextrin do not pass across the intestinal cell membrane (Wilson, 1962). When working with this substrate, one must differentiate between cavital and membrane digestion.

The involvement of the surface of the small intestine in the hydrolysis of starch has been demonstrated by several authors under conditions which, apparently, exclude other interpretation. Thus, it has been demonstrated (Ugolev, 1960a, 1960b, 1961b) that treatment of slices of the small intestines of white rats with trichloracetic acid (TCA) abolished almost completely the enzymatic activity. Furthermore, if this TCA-treated intestine is washed with buffer to remove the TCA and then placed for some time in a solution of amylase, then a certain amount of the enzyme is adsorbed on the surface of the small intestine, and the activity of the enzyme is sharply increased.† Under the conditions of these model experiments, only membrane, but not intracellular, digestion takes place.

*These methods were widely used in our laboratory and were described in the original investigations. Figures 2–4 describe the gist of these methods.

†The increase in the activity of certain enzymes, bound to membranes, compared to their activity in solution may depend on a whole series of conditions which were considered in Chapter III.

Comparable data of Lee (1962) showed that after prolonged washing in the cold of slices of small intestine of chickens to complete disappearance of the amylolytic activity, the activity of added amylase sharply increases in the presence of the small intestine.

Finally, it was recently discovered that during prolonged perfusion of the small intestine the differences between hydrolysis in vivo and in vitro gradually disappear, and after several minutes the hydrolysis of starch during perfusion is effected exclusively by the enzymes which are located in the cavity of the small intestine (Ugolev et al., 1961; De Laey and Iezuitova, 1962; Iezuitova at al., 1963; Ugolev, 1963).

From these experiments it has been concluded that amylase is gradually desorbed from the surface and for this reason the effect of membrane digestion is reduced to zero. If the perfusion is interrupted and the entry of the pancreatic juice into the intestine is retained, then after some time the membranal effects are restored. Since we are considering amylase—a somewhat heavy protein (molecular weight 45,000)—it could be assumed that the restoration of the enzymatic activity takes place as a result of adsorption of the pancreatic amylase on the surface of the small intestine.

As has been mentioned above, starch does not penetrate the intestinal cells and, therefore, it is an adequate substrate for the study of membrane digestion. Nevertheless, in investigations of this kind unexpected difficulties arose (Dahlqvist and Borgström, 1961) which indicated that the concentration of amylase in the intestinal contents is so great that the hydrolysis of starch takes place within a short time. These investigators proposed that membrane digestion does not participate in this process. This conclusion of Borgström and Dahlqvist, however, was theoretical since the authors have not performed any experiments which demonstrated that the hydrolysis of starch was entirely due to enzymes located in the cavity of the small intestine.

Iezuitova et al. (1963), Ugolev (1963, 1965b), and Iezuitova et al. (1964a) performed experiments of this nature and showed that the membranal effect increases with an increase in the perfusion rate. These data clarify the results of previous authors, including those of Dahlqvist and Borgström (1961), since in their experiments the movement of fluid in the intestine was approaching zero.

The differences in the results of those who observed high membranal effects and those who did not thus became clear. The dependence of the effects of membrane digestion on conditions of contact of the hydrolyzable substrates with the surface of the small intestine was thus discovered. The data regarding the influence of conditions of contact of the substrate with the intestinal surface permit certain important conclusions which are discussed in the section "Peristalsis of the Intestine and Membrane Digestion" (Chapter IV).

Here we should only mention that the theory of heterogeneous catalysis assumes the possibility of effective utilization of porous catalysts only during sufficiently intensive movement of fluids or gases in relation to the surface of the corresponding catalyst.

In the past years many authors have shown that in the intestine soluble starch is hydrolyzed not only by the enzymes of the intestinal contents but also by the enzymes which are adsorbed on the surface of the epithelium of the villi. Later, other data are presented which describe the role of membrane digestion in hydrolysis of starch.

HYDROLYSIS OF OLIGOSACCHARIDES

By the end of the 19th and in the beginning of the 20th century, a series of important facts had already been obtained which demonstrated that in the intestinal juice the content of oligosaccharidases is small and that these enzymes are firmly bound to the structures of the intestinal cells. Tobby and Mening (1892, cited by Florey et al., 1941) detected more invertase activity in the mucous portion than in the soluble portion of the juice obtained from the fistula of the small intestine. Bierry (1912) showed that the centrifuged supernatant of gastric juice which was passed through a bacterial filter (i.e., cell-free juice) retained only a small portion of the initial enzymatic activity. Later this was confirmed by Koskowski (1926) and Pierce et al. (1935) who showed that the intact intestinal juice is considerably more active than the centrifuged one. It was also shown that sucrose introduced into the isolated intestinal loop is split many times faster than in the juice of the same loop in vitro (Cajori, 1933).

At about the same time Bykov and Ryccle (1935) drew attention to the fact that intestinal enzymes predominate in the mucous conglomerates of intestinal juice and not in the fluid portion.

In the past few decades Shlygin (1952, 1958, 1964), Fomina (1951, 1957), and Mikhlin (1952, 1955) performed a thorough analysis of the properties of the soluble and cellular portions of intestinal juice and showed that the majority of intestinal enzymes occur predominantly in the desquamated cells. Several viewpoints were expressed on the basis of these experiments. According to one viewpoint, the final stages of the hydrolysis of oligosaccharides and other substances take place in the cavity of the small intestine by the enzymes which pass into the chyme with the destruction of desquamated intestinal cells.

According to another viewpoint, expressed by Röhmann and Nagano (1903) and Cajori (1933), the hydrolysis of oligosaccharides is effected intracellularly after the oligomers penetrate the intestinal cells.

Early in the 20th century Starling (1906) wrote that the enzymes maltase, invertase, and lactase exist primarily in the epithelial cells, since the homogenized mucous membrane of the small intestine yields an aqueous solution of these enzymes which generally is more active than the intestinal juice. In fact the difference is very large, and many physiologists assumed that the basic effect of these enzymes takes place not only in the intestinal cavity but also during the penetration of foodstuffs into the epithelial cells on the path to the bloodstream. This point of view has been supported in the past by Borgstrom et al., (1957), Borgström and Dahlqvist (1958), Dahlqvist and Borgström (1961), and others who studied digestion and absorption in the small intestine, mainly in man. They showed that oligosaccharides disappear rather quickly from the intestine, while the activity of oligosaccharidase in the intestinal contents is very low and, on the contrary, high in the homogenates of intestinal mucosa. Comparing these facts (low oligosaccharidase activity in the intestinal contents and high rate of utilization of oligosaccharides and high activity of intestinal homogenates), the authors arrived at the conclusion that cavital hydrolysis is negligible in the digestion of oligosaccharides which, according to these authors, takes place intracellularly.

The circumstance that oligosaccharides enter the medium surrounding the epithelial cells of the mucosa only in insignificant amounts has been demonstrated in experiments in vitro by Miller and Crane (1961a, 1961b), Ugolev et al. (1961), and Iezuitova et al. (1963a) on isolated slices of small intestine of rats and guinea pigs, as well as during perfusion of sections of the small intestine of rats (Iezuitova et al. 1963), and in experiments on frog intestine (Parsons and Prichard, 1965).

Apparently, it can be considered established at present that in the digestive tract only a small portion of oligosaccharides is hydrolyzed. Since convincing evidence is available to show that oligosaccharides are completely split to monosaccharides after passing the intestinal barrier (Wilson and Vincent, 1955; Chain et al., 1960; and others), it is reasonable to assume that the hydrolysis of oligosaccharides occurs somewhere in the intestinal epithelium. One cannot, however, deduce on the basis of facts proving the noncavity nature of cleavage of this or that substance that this process takes place intracellularly, because the alternative of membrane digestion exists.

In the past few years in many countries attempts have been made to determine the site of the hydrolysis of oligosaccharides. Dahlqvist

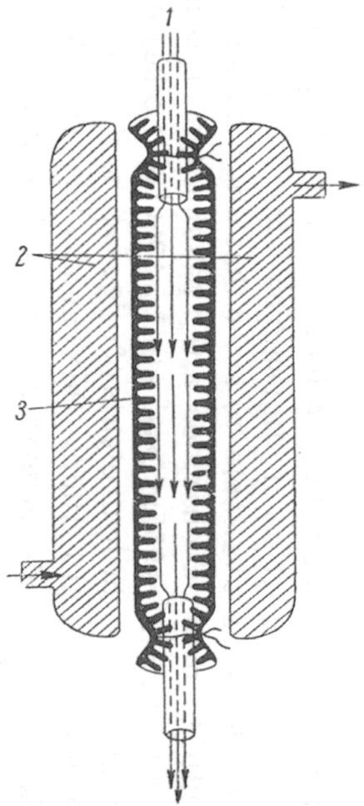

Fig. 2. Scheme of usual perfusion of an isolated segment of intestine. (1) Perfusate; (2) thermostatic muffle; (3) intestine.

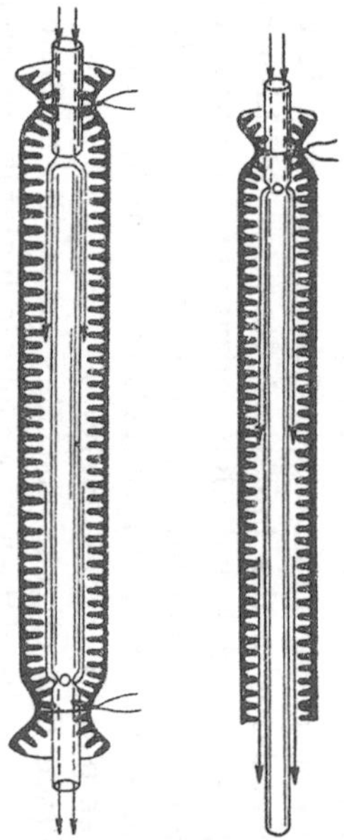

Fig. 3. Scheme of peripheral perfusion of an isolated segment of intestine which ensures best conditions of contact between the substrate and the surface of mucosa.

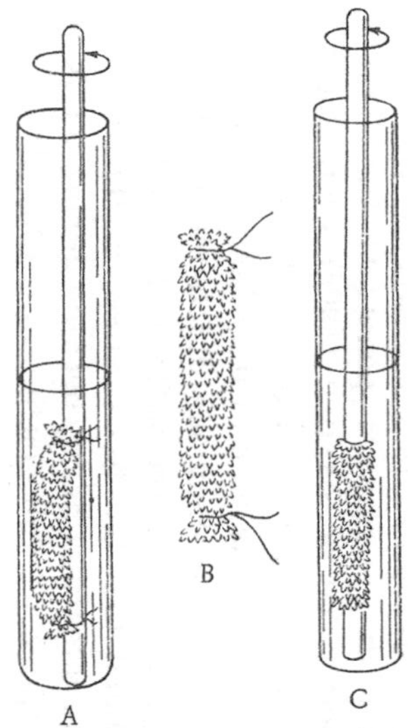

Fig. 4. Inverted sacs (A, B) and inverted slice of intestine (C) employed for study of membrane digestion. The movement of the glass rods (shown by arrows) to which the isolated segments were secured ensures intensive mixing.

(1960), employing differential centrifugation of homogenates of intestinal mucosa in sucrose medium, found that the bulk of oligosaccharidases is present in the microsomal fraction. Earlier, analogous results were obtained by Larner and Gillespie (1956) regarding maltase and oligo-1,6-glucosidase. On the basis of these data Dahlqvist concluded that the process of hydrolysis of oligosaccharides was of truly intracellular localization. Shortly thereafter, however, Miller and Crane (1961b), employing the same methods, showed that about 90% of the oligosaccharidase activity is present in the fraction of the brush border. These authors identified the subcellular structures with the aid of the electron microscope. They expressed the assumption that the data of Dahlqvist could result from destruction of the brush border, fractions of which could sediment along with the microsomes.

Later, a series of investigators employing the method of differential centrifugation confirmed that the bulk of carbohydrases sediments along with the fraction of brush border (Gallo and Treadwell, 1963; Ruttloff et al., 1964). Furthermore, Gallo and Treadwell reported that the brush border contains 96% of the invertase activity (i.e., practically the entire invertase activity). In contrast to this, enzymes involved in lipid metabolism are found in negligible amounts in the brush border. On the basis of these data one could conclude that the hydrolysis occurs somewhere near the membrane, on its external or internal surface. Understandably, this methodology cannot provide a more exact answer.

To determine the localization of oligosaccharidases more exactly Miller and Crane employed the concentration criteria of Rothstein which can be summarized as follows: if one excludes active transport and if the transfer of hexoses formed during hydrolysis of disaccharides takes place by diffusion, then one can determine the locale of the enzymatic hydrolysis from the ratio of the products of the reaction in the cell and in the surrounding medium. During the intracellular hydrolysis the concentration of hexoses will be greater in the cell than in the incubation medium, while during extracellular hydrolysis the reverse should prevail.

Miller and Crane (1961a) observed on the basis of exclusion of active transport during sucrose hydrolysis, that the concentration of hexoses in the cell was higher than in the surrounding medium. Similar data were obtained for other oligosaccharides [maltose, isomaltose, and lactose (Crane, 1962), lactose (Isselbacher and Senior, 1964)]. Newey et al. (1962, 1963) came to analogous conclusions on the basis of study of the functional topography of absorption of carbohydrates by intestinal cells.

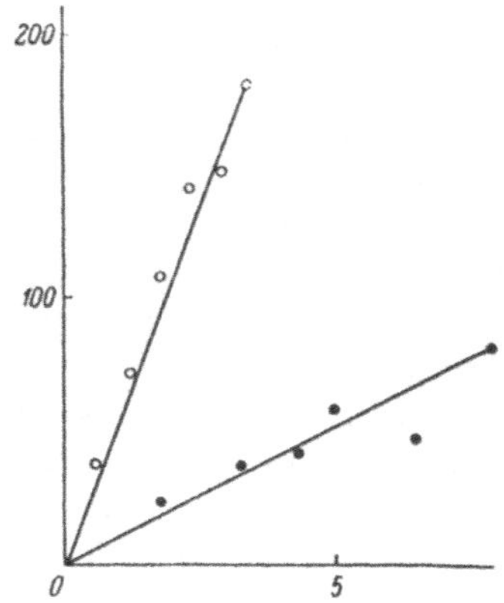

Fig. 5. Effect of the rate of perfusion on cleavage of sucrose in vivo. Perfusion of acetone-treated slices of the small intestine in a thermostatic box (two experiments). Abscissa: volume change (in ml/minute/volume of intestine in ml); ordinate: rate of hydrolysis of sucrose (in arbitrary units per minute).

Having reviewed these data we must note that the criteria of Rothstein are indisputable for the differentiation of distal extracellular and intracellular digestion, but not convincing for the differentiation of intracellular and membrane digestion. As regards the enzymatic activity of the isolated brush border, these experiments, as the authors state, cannot determine on which side of the cell membrane the hydrolytic enzymes are localized.

The problem finally became complicated when Dahlqvist and Brun (1962) showed histochemically that all epithelial cells of intestinal mucosa (from villi, crypt, and Brunner glands) contain invertase and trehalase. These enzymes were localized in small round granules scattered in cell cytoplasm. The activity of oligosaccharidases was particularly high in the perinuclear portion of the epithelial cells. No concentration of activity in the region of the brush border was noted. The authors assumed that the results of Miller and Crane were due to contamination of the brush-border fraction by the granules. It appeared to us unlikely that such an extent of contamination was present, so that 90% of the enzymatic activity resulted in the brush-border fraction.

In 1961 Ugolev (1961b) investigated the same problem. To determine the localization the oligosaccharidases, we studied the effects of the destruction of cell membranes on the rate of hydrolysis. It was assumed that should invertase be localized intracellularly the activity of the reaction should increase with the destruction of the membranes. The reverse, however, was found, and this supports the idea that invertase is located on the outer surface of the brush border. Analogous data were obtained by Dahlqvist and Thomson (1963a, 1963b).

In recent experiments by Iezuitova et al. (1963a) and by Iezuitova

(1965), the effect of movement of solutions containing sucrose in relation to intestinal mucosa on the rate of hydrolysis was investigated (Fig. 5). In these experiments the authors assumed that convection could have considerable influence on the level of formation of hexoses only if the hydrolysis takes place on the outer surface of the membrane, and only insignificant influence if it takes place intracellularly. A 2- to 10-fold increase in the hydrolytic effect because of convection was evident, and these experiments lead to the conclusion that the hydrolysis of disaccharides is localized on the external and not on the internal surface of the cell membrane.

As can be seen, almost to the present time contradictory views prevailed regarding the localization of hydrolysis of oligosaccharides in the intestine. If one takes into account, however, that the criteria of Rothstein are insufficient for the differentiation of membrane and submembrane digestion, then one can visualize the following picture. The hydrolysis takes place on the outer surface of the membrane, as documented by our experiments and confirmed by the data of Miller and Crane; Ruttloff, Frize, and Taufel; and Gallo and Treadwell regarding the localization of oligosaccharidases in the zone of the brush border. Histochemical data of Dahlqvist and Brun is understandable on the assumption, as has been demonstrated by many investigations devoted to protein synthesis in subcellular structures, that the synthesis of enzymes, including oligosaccharidases, indeed takes place in the cytoplasmic structures (ribosomes). Thus, the histochemical picture outlined by Dahlqvist and Brun could characterize the location of synthesis but not of the activity of oligosaccharidases.

Membranal localization of the activity of oligosaccharidases received considerable support in recent studies by Lindemann and Solomon (1962) who showed that the effective radius of pores of membranes of intestinal epithelium is about 4 Å (as in other tissues). Thus, the molecules of substances with a large radius and which are insoluble in lipids (including sucrose) do not penetrate intestinal cells.*

HYDROLYSIS OF PROTEINS AND LARGE PEPTIDES

In adult animals the intestinal wall is a barrier which is impenetrable to proteins and large split products of protein molecules.

*With the aid of the immunochemical technique the presence of β-galactosidase in the brush border has been demonstrated (Doell et al., 1965). This, along with other data, leads to the conclusion that the hydrolysis of lactose takes place in the region of membranal digestion.

This problem has been reviewed thoroughly by London (1916, 1938), Brambell (1958), Faitelberg (1960), Wilson (1962, 1965), and Saunders and Isselbacher (1966). On the contrary, the prenatal and early postnatal periods of development of mammals is characterized by high penetrability of the intestinal barrier by large molecules, including proteins. It is possible that the basic mechanism of this transfer is the well-demonstrated pinocytosis. Clark (1961a, 1961b) and Wilson (1962) described a strict correlation between the sharp decline in pinocytotic activity of intestinal cells and the cessation of entry of proteins from the intestine into the internal medium. This is an important observation, inasmuch as during the investigations of the mechanism of protein hydrolysis in adult animals one must differentiate between cavital and membrane digestion.

It is generally considered that proteins are subjected to prolonged action by stomach and pancreatic enzymes in the intestinal cavity. Were it so, then after the introduction of amino acids and protein hydrolysates into the intestinal canal a higher rate of absorption of the latter should prevail, compared to that after the ingestion of proteins from which the hydrolysates were obtained.

However, a large number of recent investigations has shown that the cleavage of proteins in the intestinal canal and the absorption of the end products take place considerably faster than one can expect from theoretical considerations. Gupta and others (1958) investigated the intestinal absorption as follows: rats were administered test breakfasts consisting of 75% sucrose, 10% vegetable oil, and 15% protein in one series of experiments and 15% of a mixture of amino acids in others. One could expect that the nitrogen of the amino acids would be absorbed faster than the nitrogen of the protein. However, the authors could not detect any significant difference in the rate of removal of food from the intestine under both conditions.

Crane and Neuberger (1960) studied the rate of digestion and absorption of proteins in healthy humans. The rate of absorption was measured by the appearance of the N^{15} label in the urine. In one series of experiments N^{15}-labeled yeast protein was used; in the others, a hydrolysate of the same protein obtained after prolonged digestion with pepsin, then with pancreatin. Tests have shown that the hydrolysate contained 61% cleaved peptide bonds. Under these conditions one could expect the hydrolysate to be absorbed faster than the undigested protein. However, the intact yeast protein and its hydrolysate were absorbed at about the same rate.

The question arises regarding the mechanism which acts in vivo and which ensures a very rapid hydrolysis of proteins to oligopeptides and amino acids. Experiments conducted by Ugolev et al. (1961) and

Ugolev (1963) have shown that the hydrolysis in the small intestine of white rats occurs considerably faster than that by the enzymes present in the intestinal contents. The activity of the latter was determined in vitro. It turned out that during perfusion of the upper portions of the small intestines of rats by denatured blood serum, the hydrolysis in vivo was on the average 2 to 2.5 times faster than the hydrolysis by the enzymes which entered the perfusate. This was observed also in experiments in which the entry of pancreatic juice into the small intestine was preserved as well as in those in which the entry of pancreatic juice into the intestines was prevented.

Since proteins cannot penetrate the intestinal barrier, the high rate of hydrolysis in vivo can be explained only by assumption that the activity of enzymes within the intestines takes place concurrently with the activity of enzymes which are adsorbed on the outer surface of the cell membranes. In many ways these experiments are similar to those in which studies were made regarding the effects of intestinal surface in the hydrolysis of starch. The similarity was all the more evident by the fact that during prolonged perfusion the dissimilarity between the hydrolysis of proteins in vivo and in vitro gradually disappeared. This, apparently, was connected with the gradual desorption of enzymes from the intestinal surface.

Thus, during the hydrolysis of proteins as well, the rate of cleavage in vivo is greater than could be expected on the basis of enzymatic activity of the intestinal contents. The catalyzing mechanism is provided by membrane digestion which takes place on the outer surface of the intestinal epithelium. The hypothesis of intracellular digestion appears to be improbable in this case because of the impenetrability of the intestine of adult animals to polypeptides.

HYDROLYSIS OF OLIGOPEPTIDES

The history and present status of the question of the localization of the final stages of hydrolysis of peptides are in many ways similar to the history of the question of hydrolysis of oligosaccharides. This is understandable since in both cases the point is hydrolysis of oligomeres which could, according to several investigators, penetrate the cells.

The fact that the cleavage of peptides to amino acids does not take place in the cavity of the small intestine was first realized by London (1916). In his major investigations of chyme composition in dogs with polyfistulas, London showed that the concentration of amino acids in the intestinal contents is lower than could be expected if the hydrolysis of protein to amino acids takes place in the intestinal cavity. Since an increased content not only of amino acids but of peptides as well was observed in the portal blood of angiostomized dogs London and Lovitskii (1938) proposed that peptides can enter the blood without hydrolysis and serve as the building blocks of proteins. This viewpoint has been lately supported by Fisher (1954).

Cajori (1933) studied the rate of absorption of peptones in isolated intestinal loops of dogs and found that the rate of absorption was several times higher than the peptidase activity of the intestinal juice. He, as had London, concluded that the cells of intestinal epithelium can resorb not only amino acids but peptides as well. On the basis of previous investigations, however, he assumed that inside intestinal cells, cleavage of peptides to amino acids takes place and these enter the blood.

In the past ten years this problem once more was subjected to thorough investigation employing newer technical devices. It has been shown that at least small peptides can pass in small amounts across the intestinal barrier into the blood (in experiments in vivo or when incubated in vitro) (Agar et al., 1953, 1954; Newey and Smyth, 1957, 1959, 1960a, 1960b; Johnston and Wiggans, 1958; Wiggans and Johnston, 1959; see review by Wilson, 1962). On the other hand, it has been established that the bulk of the peptides leave the intestine as free amino acids (Parshin and Rubel, 1951; Yuile et al., 1952; Agar et al., 1953; Denton et al., 1953; Denton and Elvehjem, 1954; Hartmann et al., 1956; Levenson et al., 1959). Some authors categorically deny the possibility of penetration of peptides across the intestinal barrier. Parshin and Rubel (1951) have shown in chronic experiments on dogs and in acute experiments on cats that after feeding the animals meat, casein, and liver the level of amino acids in the blood rises. Absorption of peptides was not observed. Levenson et al. (1959) fed dogs liver and homologous serum protein of dogs by means of catheters placed in mesenteric veins. Analyses of blood taken from these veins showed that the amino acid levels six hours after the feeding had increased. Peptides were not detected. According to these authors, homologous as well as heterologous proteins are hydrolyzed completely to amino acids. Thus, the most important conclusion to be drawn from these data is the intensity of the hydrolysis of peptides in the intestine.

For the elucidation of the localization of the hydrolysis of peptides, isolated sacs (the regular and particularly the inverted) proved very useful (Wilson and Wiseman, 1954; Wilson, 1962). During incubation of these preparations it is possible to study all stages of transfer of substances across the intestinal wall. This technique has been extensively used in recent work, particularly in comprehensive and well-known experiments of Newey and Smyth (1957, 1959, 1960a, 1960b,

1961, 1963), Johnston and Wiggans (1958), and Wiggans and Johnston, 1959). According to their data, peptides disappear uncleaved from the mucous side and appear on the serous side as amino acids (except glycylglycine which is transported partially uncleaved). Of considerable interest are the data of Wiggans and Johnston (1959) who followed the cleavage and transport of several dipeptides. On the basis of these experiments a conclusion has been reached that peptides are hydrolyzed intracellularly. This conclusion is of importance and it appears appropriate to examine the basic considerations of the authors.

During the incubation of the isolated sacs, the peptide is initially only in the solution which is in contact with the mucosa. One hour after the initiation of incubation, the mucous as well as serous solutions contain considerable amounts of amino acids. According to Newey and Smyth, hydrolysis of a peptide could be determined by: (1) peptidases which are present in solutions in contact with the mucosa; (2) peptidases which are present inside the cells of the mucosa and which act on the peptides entering these cells; and (3) peptidases which are present in serous fluid and which act on peptides which escaped intracellular hydrolysis and entered the serous fluid.

Since the peptidase activity of fluids in contact with the isolated intestinal sac from both sides is extremely small, the conclusions regarding the intracellular hydrolysis of peptides appear reasonable.

We should like to note, however, that these excellent experiments, which showed with precision that hydrolysis of peptides is induced by enzymes bound to cellular structures, do not include approaches which could permit the differentiation between the intracellular and membrane digestion, but permit with equal certainty the assumption of either mechanism. Since Newey and Smyth observed that even during intact active transport of amino acids a considerable amount of end products of hydrolysis of peptides is present in the fluid around the mucosa, these authors assumed that the amino acids which are formed during the intracellular hydrolysis are in part once again returned to the fluid surrounding the mucosa, while the other portion is transported across the intestinal wall (Newey and Smyth, 1960a, 1960b; Smyth, 1961).

These data are more in accord with the assumption of hydrolysis on the outer surface of cellular membranes, while the presence of amino acids in the fluid surrounding the mucosa can be considered as a result of the fact that the rate of absorption is somewhat slower than that of hydrolysis.

Histochemical investigations of its aminopeptidase activity were performed almost simultaneously with the study of enzymatic functions of the small intestine (Nachlas et al., 1960; Burstone and Weisburger,

1961; see also review by Burstone, 1965). It was found that aminopeptidase is localized in the region of the free intestinal surface. Analogous data were then obtained by Holt and Miller (1962) who isolated the brush border by means of differential centrifugation and showed that leucine aminopeptidase is predominantly localized (as is alkaline phosphatase) in the brush border. It is understandable that the latter two facts do not permit the solution of the problem of on which side of the cellular membrane the hydrolysis takes place.

The data obtained in our laboratory basically duplicate the data of Newey and Smyth and Johnston and Wiggans. For reasons given above, they also do not permit favoring either intracellular or membrane digestion (Ugolev et al., 1961; Ugolev, 1963).

As noted above, histochemical approaches and the isolation of the brush border permitted the assertion that the peptidase activity is localized in the region of the free intestinal surface. In order to determine whether this process takes place on the inner or the outer side of the membrane, we investigated the effect of movement of the fluid in relation to the surface of the mucous membrane on the intensity of peptide hydrolysis. A sharp increase in the cleavage of glycylleucine and alanylserine indicates that at least these peptides are cleaved predominantly on the outer surface of the cellular membrane, although this method does not permit complete exclusion of the role of intracellular digestion (Fig. 6).

HYDROLYSIS OF FATTY ACID ESTERS

The mechanism of hydrolysis of glycerol esters of fatty acids has been described in detail in several recent reviews (Spencer, 1960; Isselbacher, 1961; Ugolev, 1961a, 1963; Wilson, 1962; Clément, 1964;

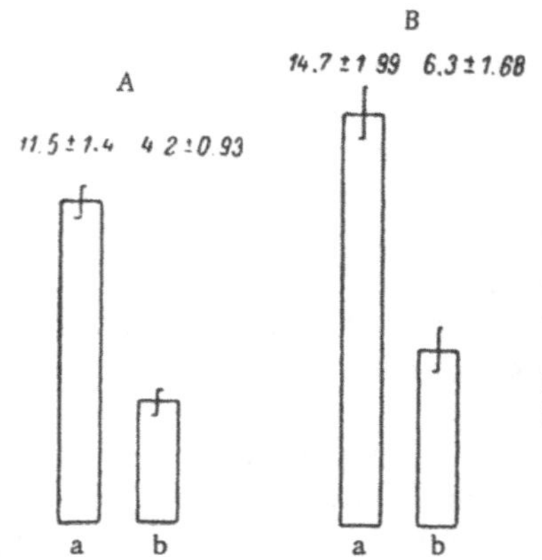

Fig. 6. Hydrolysis of dipeptides—glycyl-L-leucine (A) and DL-alanyl-DL-serine (B)—during mixing (a) and without mixing (b). Numbers indicate the increase in free amino nitrogen (in % of hydrolyzed dipeptide in 1 minute; average value ± standard error).

and Isselbacher and Senior, 1964) and in a symposium devoted to lipid metabolism (Frazer, 1963).

It is considered established at present that the hydrolysis of esters of fatty acids takes place only partially in the gastrointestinal tract; the extent of this hydrolysis and the chemical state in which fats pass across the membrane surface of the intestinal epithelium are as yet debatable. Considerable advance in the investigation of these problems have been possible since about 1950 because of the introduction of several new methods of investigation of digestion and absorption of fats which are listed in the review by Clément (1964).

Present concepts can be summarized as follows. In the intestine, fatty acids are stepwise cleaved from the molecule of triglyceride leading to stepwise formation of di- and monoglycerides and free fatty acids.

As early as 1945 Frazer and Sammons (1945) showed that during in vitro digestion of fats by pancreatic lipase very little glycerol was formed in the first few hours, whereas monoglycerides were formed in considerable amounts. Desnuelle et al. (1947, 1948) in experiments in vivo found that the hydrolysis of glycerides takes place stepwise and the cleavage of free acids takes place with increasing difficulty; thus, diglycerides are easily formed while the monoglycerides are formed much less so.

It has been established that bile salts favor the hydrolysis of fats, although the mechanism of their action is far from clear. It is possible that the addition of bile increases the intensity of hydrolysis of fats by raising the optimum pH of lipase to values which prevail in the upper portions of the small intestine (Borgström, 1954). An assumption has been expressed according to which the role of bile acids consists in the alteration of physicochemical properties of the enzyme–substrate complex. It has been shown that mixed substances, namely bile salts and monoglycerides, are cleaved much more easily by pancreatic lipase (Hofmann and Borgström, 1963). These data were confirmed by Clément (1964).

In a series of investigations in vivo by various authors it has been found that, as a rule, during digestion the intestinal contents produce a mixture which consists of considerable amounts of triglycerides, smaller amounts of diglycerides, and comparatively small amounts of monoglycerides (Desnuelle et al., 1951; Borgström, 1952; Desnuelle and Constantin, 1952; Kuhrt et al., 1952; Harris et al., 1955; Knoebel and Nasset, 1957).

Data are available regarding the specificity of lipase action. Thus, during hydrolysis the fatty acids attached to glycerine in positions 1 and 3 are cleaved off (Mattson et al., 1952).

Regarding substrate specificity, it has been shown that pancreatic lipase selectively acts on the triglycerides of long chain fatty acids which are insoluble in water (Ravin and Seligman, 1953; Sarda and Desnuelle, 1958).

It has been established that the hydrolysis of triglycerides containing short chain fatty acids proceeds faster than the hydrolysis of triglycerides of long chain fatty acids. Data are available (Savary and Desnuelle, 1956) regarding the similarity in the rates of hydrolysis between C_{12} to C_{18} fatty acids present in the triglycerides.

Lipolysis of triglycerides with short chain fatty acids (C_2 to C_{10}) proceeds more intensively the shorter the fatty acid chain (Entressangles et al., 1961; Clément et al., 1962). Free fatty acids, mono-, di-, and triglycerides can enter the intestinal cell (Fig. 7).

Consideration of the metabolism of fat within the intestinal cell is outside the limits of the present book. We should mention, however, that data are available that indicate the possibility of resynthesis of fats not in the cavity but only during the passage across the epithelial cells of the mucosa of the intestine. This fact has been confirmed by many authors employing simultaneously introduced free labeled glycerol and glycerides or free fatty acids, followed by the determination of radioactivity of glycerides in the lymph (Favarger and Collet, 1950; Favarger et al., 1951; Bernhard et al., 1952; Buensod et al.,

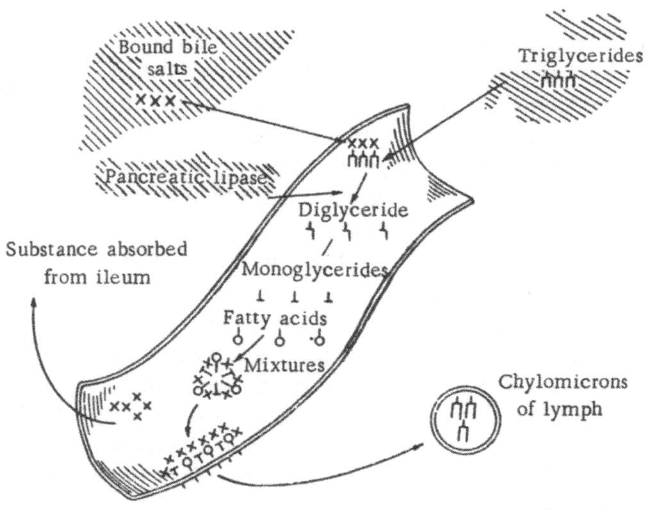

Fig. 7. Schematic representation of certain processes taking place in the small intestinal canal which are of significance to digestion and absorption of fats (according to Isselbacher and Senior, 1964, p. 293, Fig. 2).

1953). Isselbacher and Senior (1964) in their review list several data regarding the fact that hydrolysis of monoglycerides is effected by enzymes which are synthesized by the intestinal cells. Of considerable interest are the data of Mohiuddin (1964) who showed that pinocytosis in the intestinal cells of adult animals cannot play any significant role in the hydrolysis and utilization of triglycerides.

Clément (1964) in his major review concludes that the quantity of intact triglycerides during absorption does not exceed 10% and that 80% of the fatty acids of the ingested triglycerides are in a free state at the end of digestion. The remaining 10% are in the form of mono- and diglycerides.

The basic sites of absorption of fats are the upper sections of the intestine, namely the end of the duodenum and 50–100 cm of the jejunum (Borgström et al., 1963).

Ugolev (1960b, 1961a, 1963) showed that lipases play a definite role in the hydrolysis of triglycerides being fixed on the outer surface of cell membranes, particularly the pancreatic lipase which is adsorbed on the surface of the intestinal epithelium from chyme.

A detailed study of the role of the intestinal surface in the hydrolysis of fats has been made by Nurkes and Ugolev. They showed that inverted surviving sections of small intestine, kept during repeated incubations in tributyrin solutions, gradually lose their activity; that is, the tributyrinase is apparently desorbed from the surface as amylase is. In all tests, the activity in the presence of the intestine was several times higher than in its absence, indicating the presence of a considerable degree of hydrolysis on the intestine surface.

It is of interest that, unlike amylase, whose activity on the surface falls off during repeated incubations, tributyrinase activity of the intestinal surface is preserved even during prolonged desorption. This permits the conclusion that tributyrinase activity is only partially determined by the adsorption of pancreatic lipase on intestinal epithelium, while it is partially determined by enzymes of the intestine itself which possess the tributyrinase activity (Fig. 8).

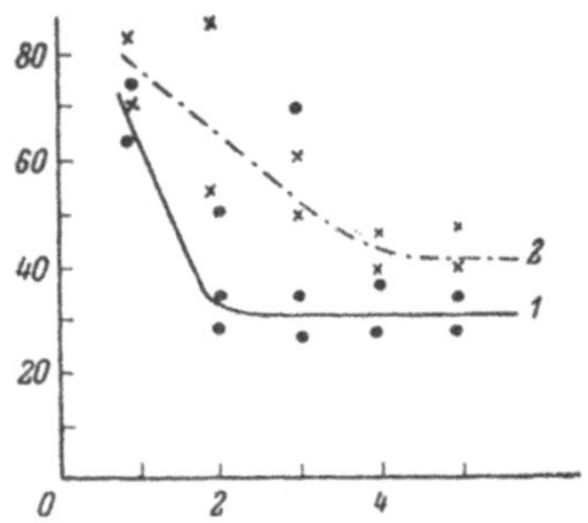

Fig. 8. Hydrolysis of tributyrin by intestinal slices during repeated incubations. Abscissa: repeated incubations; ordinate: tributyrinase activity (in % of hydrolyzed tributyrin). (1) Acetone-treated intestine; (2) intestine not treated with acetone.

In the same investigations, clearly defined membranal effects were detected in the frontal portion of the small intestine. The effects in the third and fourth quarters of the intestine were considerably smaller. Analogous data were obtained also during peripheral perfusion of isolated sections of the small intestine with tributyrin solutions. It has been shown histochemically that in the region of the brush border esterase and lipase were present. Biempica et al. (1961) observed the predominant activity of esterase and lipase in the brush-border region of the mucous membrane of the small intestine obtained from healthy humans by aspiration biopsy.

As mentioned earlier, histochemical investigations which are based on light microscopy do not permit the solution of the problem of membrane or intracellular digestion. In our work, however, the possibility of desorption of lipase from the surface of the small intestine has been indicated; it is possible only if the enzyme is localized on the outer surface of the cell membranes.

Thus, membrane digestion may play a definite role in the digestion of fats. This, however, requires further detailed study.

HYDROLYSIS OF PHOSPHORIC ACID ESTERS

Hydrolysis of phosphoric acid esters is accomplished by several enzymes. A group of these is known in the literature collectively as alkaline phosphatase. In reality, to this group belong enzymes which hydrolyze phosphoric esters in alkaline medium (hydrolysis of ATP, hexose phosphates, glycerophosphates, cleavage of phosphoric acid nucleotides, and synthetic substrates such as napththol phosphate, phenolphthalein phosphate, etc.). Since several enzymes which belong to the group of alkaline phosphatase possess general but not absolute specificity, it is difficult to ascertain, employing one of the substrates, precisely what a given enzyme is accomplishing under natural conditions. The localization of alkaline phosphatase has been studied in the small intestine of higher animals and man by a variety of methods. Particularly numerous are the histochemical studies. Following Gomori (1939, 1941) who developed the most widely used method, many authors observed that alkaline phosphatase is localized mainly in the region of the brush border (Bourne, 1943; Deane and Dempsey, 1945; Dempsey and Deane, 1946; Emmel, 1946; Martin and Jacoby, 1949; and many others).

Even by that time it had been shown that the histochemical localization (distribution) of alkaline phosphatase depends on the length of incubation of the tissue with the substrate. During more prolonged incubations, a diffusion of the reagent inside the intestinal cells was observed; this led earlier investigators to an erroneous conclusion regarding a uniform distribution of alkaline phosphatase throughout the cytoplasm of the cell. During shorter incubations alkaline phosphatase was practically entirely concentrated in the region of the brush border.

As we mentioned earlier, the concept of alkaline phosphatase embodies a whole group of enzymes. Certain contradictions in the experiments depend on what substrate and, therefore, what enzyme, is being investigated. Dempsey and Deane (1946) made very interesting histochemical studies of the localization of phosphorylation processes of various substrates. They established that fructosediphosphatase at pH 7.0 is detected in the brush border as well as in the region of the Golgy apparatus and also at pH 4.5 in the nuclei. The distribution of phosphatase of nucleic acids was approximately similar.

Dephosphorylation of glucose-1-phosphate takes place in the brush border as well as in the nuclei, cytoplasm, and *muscularis mucosae*. A positive lecithine-phosphatase reaction is detectable in the brush border and in nuclei. Hydrolysis of glycerophosphate is always observed in the brush border and, depending on changes in pH, in certain other cell structures.

Thus, as early as the 40's, it was shown that alkaline phosphatase is concentrated in the region of the cell surface. Combining histochemical approaches with an electron microscopy permitted an even more exact locating of alkaline phosphatase in the intestinal cells (Barrnett, 1959; Zetterqvist and Hendrix, 1960; Clark, 1961a, 1961b).

With all certainty it can be stated that phosphatase is located in the cell membrane. The fact that after short incubations the stain is revealed on the surface of the microvilli and its diffusion into the cytoplasm is not observed can be considered as proof of the location of the enzyme on the surface (Barrnett, 1959).

Analogous data were obtained during studies of the distribution of ATP-ase in the brush border of the small intestine of rats (Ashworth et al., 1963). A photomicrograph (Fig. 9) shows that the product of the reaction is located in the spaces between the almost-vertically sliced microvilli. The same conclusion can be drawn from the data of Rothstein et al. (1953) who studied the hydrolysis of P^{32}-labeled glucose-1-phosphate. As is known, this substrate penetrates across the cell membranes, and its intensive hydrolysis in the intestine as well as the distribution of phosphate and glucose during hydrolysis indicates that the latter occurs at the outer surface of the cells. Apparently, the pyrimidine nucleotides are cleaved at the same site (Wilson and

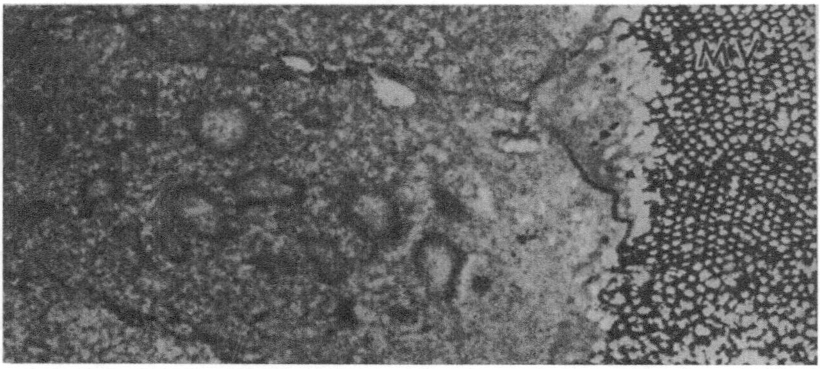

Fig. 9. ATP-ase reaction in the mucosa of the small intestine. ×10,000 (reduced 15% for reproduction) (according to Ashworth et al., 1963, p. 9, Fig. 10). The products of the reaction surround the cross-cut microvilli (MV).

Wilson, 1958). The data of Rothstein et al. are well in accord with the observations of Holt and Miller (1961, 1962) who showed that during differential centrifugation of the intestinal mucosa homogenate the bulk of phosphomonoesterase sedimented together with the brush-border fraction.

Thus, studies using several methods have shown the localization of a series of dephosphorylating enzymes in the region of the brush border of intestinal cells; in some cases it was found that the enzyme is fixed at the outer surface of the membrane (additional data on this are considered in subsequent chapters).

However, a generalization of these data is difficult since alkaline phosphatase activity was also detected in tissues which definitely are not involved in digestive processes, namely renal canaliculi and erythrocytes (see review by Rothstein, 1954).

The often observed similarity in the localization of alkaline phosphatase in intestinal epithelium, renal canaliculi, erythrocytes, and secretory organs suggests that the phosphatases of the cells perform certain important functions connected with active transport. Keeping in mind the localization of alkaline phosphatase of the kidneys and secretory organs on the surface, the viewpoint expressed by Bradfield (1950), Danielli (1952), Rosenberg and Vilbrandt (1955) and many others is particularly logical, namely that alkaline phosphatases are enzymes which are involved in the active transport across the membranes. True, certain authors (see reviews by Crane, 1960, 1962) deny the importance of processes of dephosphorylation in glucose transport; however, there are other viewpoints (Shishova et al., 1961; Wilson, 1962). Thus, ample data are available which favor the view that alkaline phosphatases are present in the membrane, particularly its surface. In most cases, the role of these enzymes is not clear. It is possible that they are connected with the active transport across the cell membrane. It can be assumed that at least in the small intestine certain alkaline phosphatases are functioning as digestive enzymes.

* * *

We attempted to present a traditional review of the hydrolysis of basic foodstuffs in the small intestine of higher animals. As can be seen, the classic view of cleavage of foodstuffs taking place entirely in the gastrointestinal tract has been succeeded by other views. Already in the first decades of the 20th century attention had been focused on the importance of intestinal epithelium in the performance of the final stages of hydrolysis. The available data in this direction were interpreted in terms of the other type of digestion then known, namely intracellular.

In the past few years a series of facts have been obtained indicating the existence of membrane digestion. Thus, the elucidation of which type of digestion is responsible for the hydrolysis of this or that foodstuff substrate is often a difficult problem. In addition, the reader can see that often the authors arrive at different conclusions on the basis of identical results. This is true, in particular, concerning the question of the membrane and intracellular mechanism of digestion of peptides and oligosaccharides.

We will now review the criteria which were employed in the resolution of the problem of localization of the hydrolytic processes and attempt to evaluate the conclusions which can be made employing appropriate methods of investigation.

CRITERIA EMPLOYED IN THE INVESTIGATION OF LOCALIZATION OF DIGESTIVE PROCESSES

Differentiation of Distal Extracellular Digestion from Cellular Digestion (Intracellular and Membrane)

In the majority of cases it is comparatively easy to differentiate between the hydrolytic processes dependent on enzymes bound to cell structures from those which are performed by enzymes in the surrounding medium (this could be the medium in which the organism exists or the contents of the intestinal tract). If the substance under investigation is cleaved but the corresponding enzymes in the medium surrounding the cell are absent or are present in significant amounts, one can conclude that membrane or intracellular digestion is present. If the cleavage of substrate takes place in the medium containing the corresponding enzymes, then distal digestion is evidently taking place.

However, even in the latter case it is erroneous to deny the existence of either type of cellular digestion. To solve the problem of whether or not distal hydrolysis under given conditions is the sole mechanism it is imperative to compare the rates of hydrolysis in the presence or absence of cellular material.

Thus, for instance, the rate of the hydrolysis of proteins in the stomach and in vitro by the enzymes of stomach contents is about the same, and this permits the conclusion that in this organ cavital digestion takes place practically exclusively. The rate of the hydrolysis of proteins and polysaccharides in the intestinal tract is considerably higher than that performed in vitro by the enzymes of the contents of the intestine. Here one must arrive at the conclusion that the hydrolysis in vivo is accomplished by the enzymes acting in the cavity of the small intestine as well as by the enzymes bound to the structures of the intestinal cells.

It appears important to direct attention once more to the circumstance that the presence of enzymatic activity in the contents of digestive cavities or in the fluid surrounding the cellular material, indicating the presence of distal digestion, does not offer adequate grounds for the denial of the presence of cellular digestion. Final conclusions are made on the basis of comparative rates of the reaction in vivo (that is, in the presence of cellular material or after the introduction of the substance into the appropriate digestive cavity) and of the reaction in vitro (by the enzymes which enter the incubation medium or which enter the appropriate digestive cavity).

In general, it is comparatively easy to differentiate between distal and cellular digestion. The greatest difficulties arise when it becomes necessary to differentiate membrane from various types of intracellular digestion. We shall, therefore, dwell in detail on this problem.

Differentiation of Membrane from Intracellular Digestion

1. If the substrate is not hydrolyzed in the medium surrounding the mucosa and does not penetrate the cell, then the cleavage of the substrate in the presence of cellular material (with uninjured membranes) is due, apparently, to membrane hydrolysis.

2. If enzymatic inhibitors do not penetrate the cell and if in their presence the enzymatic activity is sharply reduced or absent, then the conclusion that the hydrolysis was membranal and not intracellular appears very probable. One must, however, keep in mind that it is theoretically possible to deal in this case with the alteration of true

intracellular hydrolysis because of a blocking by the inhibitors of the passages or pores along which the foodstuff substrates penetrate the cell. This criterion was employed with great success in the demonstration of the existence of membrane digestion in yeast, when it was shown that the treatment of yeast cultures by heavy metal ions which do not penetrate the cell prevents the hydrolysis of sucrose. It follows that the absence of the effect during the action of inhibitors which do not penetrate the cell membrane can serve as proof of true intracellular digestion.

3. If the activity of enzymes is altered with the changes in pH of the medium containing uninjured tissue or cell culture, then it is reasonable to assume membrane digestion. Relative independence of enzymatic processes from the pH of the medium, on the other hand, suggests intracellular digestion. This criterion was also employed in the studies of the site of hydrolytic processes in yeasts.

In the utilization of this criterion erroneous conclusions could be made if one does not take into consideration the particular relationship between the adsorbed enzymes (compared to dissolved enzymes) and the pH of the medium. In addition, the pH of the medium, while not affecting the enzymatic activity, may sharply alter the permeability of the membranes toward various substances (see review by Wilson, 1962).

4. Study of the equivalent radii of membranal pores of various cells and comparison of these with the size of the molecules could be an argument in determining the site of digestive processes. Thus, if the substrate molecules are greater than the radius of the appropriate pores, it is difficult to assume that the hydrolysis takes place intracellularly.

5. Of considerable importance is the study of the distribution of enzymatic activity in the cell by histochemical procedures. If by the use of these methods the enzyme is shown to be evenly distributed in the cell cytoplasm, then it could be assumed that one is dealing with intracellular digestion. Such a conclusion appears even more probable if the enzyme is shown to be bound to definite cell organoids.

With such data at hand, however, one must keep in mind three complicating circumstances. (a) The true intracellular localization of an enzyme may reflect not the site of its activity but the site of synthesis of the investigated material. (b) Not infrequently intracellular enzymes are detected in cells which definitely do not participate in intracellular digestion, for instance, invertase in stomach cells (Dahlqvist and Brun, 1962). (c) The history of investigation of the localization of alkaline phosphatase convinces one that the most suitable conditions for the establishment of enzymatic activity are very short periods of incubation and adherence to rigid measures to prevent diffusion. Indeed, it first appeared that alkaline phosphatase is present throughout the intestinal cell; on improvement of technique it was found that the enzyme was basically concentrated in the region of the brush border (for details see the section "Hydrolysis of Phosphoric Acid Esters," p. 26).

In those cases when the enzyme is shown to be localized in the region of the free cell surface, one should consider the possibility that the digestion may take place on the outer surface of the cell membrane as well as on the submembranal surface (intracellularly). Classic histochemical methods cannot differentiate between these processes.

6. Methods which combine histochemistry with electron microscopy for the differentiation of intracellular from membrane digestion appear to be very promising. It was by such methods that it was possible to show that the alkaline phosphatase of the small intestine of animals is present on the outer and not on the inner surface of the membrane (Brandes et al., 1956; Barrnett, 1959; Clark, 1961a, 1961b).

7. The study of the localization of digestive enzymes in the cell can be facilitated by the use of preparative separation of subcellular structures. For this purpose the method of differential centrifugation in a sucrose medium was successfully employed. If the bulk of the enzyme under study is concentrated in the cytoplasmic structures or in the soluble fraction, then it could be concluded that the enzyme acts intracellularly (with the same reservations as those made for histochemical methods). If the enzyme accompanies the membrane fraction, then, in these cases, as in the histochemical studies, one may suppose either submembranal or extracellular digestion, and additional methods are required for differentiating between these processes.

8. The concentration criterion received wide application, and it was successfully employed in the study of the localization of yeast invertase by Rothstein (1954). It later was employed in the study of the localization of oligosaccharidases in the small intestine of mammals. Let us assume the existence of two compartments, *a* and *b*, which are divided by a membrane permeable to dissolved substances. The problem is to determine in which of the two compartments the conversion of a polymer to a monomer takes place. Such a problem is easily resolved by a comparison of the concentrations of the reaction product in both compartments. Indeed, if the concentration of the reaction products is greater in compartment *a* than in *b*, then it could be deduced that the reaction takes place in compartment *a* from which, in accordance with the diffusion gradient, the products pass into compartment *b*. To obtain a reliable answer, two conditions appear to be necessary: (1) one must exclude active transport against the concentration gradient and (2) utilization of the substances under study must be excluded.

If active transport and utilization of reaction products can be excluded by means of appropriate inhibitors, then the described model may be employed for determining the site of the hydrolytic processes. Indeed, upon comparison of the concentrations of reaction products inside the cell and in the surrounding medium if one finds that the concentration of the substances under study is greater in the surrounding medium, then one may assume extracellular hydrolysis; if the reverse is true, intracellular hydrolysis is evident.

Unfortunately, this criterion, which only recently appeared to be absolutely reliable, may lead to equivocal conclusions in cases when hydrolysis is localized on the membrane which divides the intracellular and extracellular fluids. Indeed, in such cases the concentration maximum will be at the cell surface from which diffusion is possible either inside the cell or into the surrounding medium, and the ratio of the concentrations will depend on a series of circumstances, particularly on which side of the membrane hydrolysis takes place, on the permeability of the membrane, and, finally, on the capacity of the nuclei located on both sides of the membrane. In the case of hydrolysis on the cell surface it is theoretically possible that higher concentrations will prevail on either side of the dividing membrane.

Thus, the concentration criterion becomes absolute in the differentiation of true intracellular and extracellular distal digestion, but it may lead to erroneous conclusions when one attempts to employ it to differentiate between membrane and intracellular digestion.

9. If one deals with a multistage reaction, then, as a rule, the rate of the reaction as a whole is limited by the slowest stage. In the case of intracellular hydrolysis such a stage is the passage of oligomers inside the cell across the membrane; in the case of membrane digestion it is the diffusion of initial substances involved in the reaction from inside the phase to the surface (Germain, 1961). It is therefore clear that in the case of intracellular hydrolysis the movement of the fluid in relation to the cell surface should not be reflected in the rate of penetration of substances across the membrane, since the rate of surface processes within wide limits is proportional to the rate of movement of the fluid in relation to the catalytic surface (Fig. 10). This criterion may be employed for the differentiation of intracellular and membrane digestion; it requires, however, the use of special controls which permit one to determine to what extent mixing affects the rate of penetration of substances under study or substances related to them inside the cell.

This does not exhaust the list of criteria which could be useful for the differentiation of various types of digestion. Indeed, the list is growing, and later we describe in detail certain new criteria whose

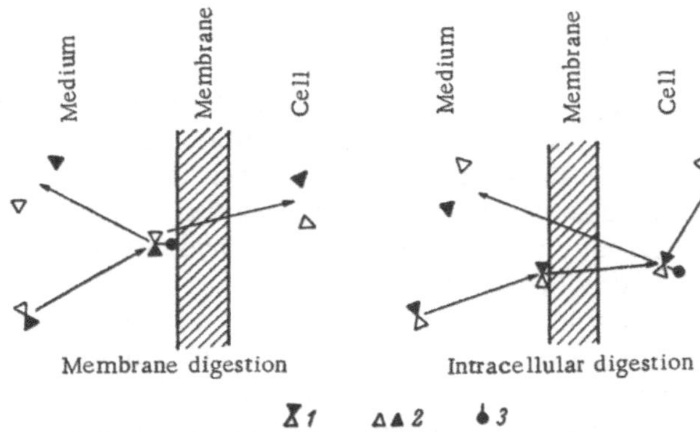

Fig. 10. Essential scheme of experiments on the effect of mixing on hydrolysis of dimers. The arrows point to movement of di- and monomers. (1) Dimer; (2) monomer; (3) enzyme.

use proved advantageous in the elucidation of the true localization of digestive processes in the small intestine of certain mammals. The most important conclusion is that a clear understanding of the resolving possibilities of the methods employed is essential. Clearly, underestimation of this may lead to erroneous deductions. On the other hand, a clear understanding of the possibilities offered by a method may save the science the frequent confusion of interesting speculations with direct deductions.

From this standpoint, let us examine the previously mentioned data on the mechanism of digestion in higher animals.

Apparently, all available data to date could be summarized in three groups. The first group comprises observations which indicate that the cleavage of oligomers is induced by enzymes which are bound to structures of the intestinal cells. The facts related to this group do not permit the solution of the question as to which type of digestion (membrane or intracellular) is taking place.

The second group of facts indicates that the hydrolysis takes place in cytoplasmic structures or on the inner surface of the membrane, that is, intracellularly. Let us examine these data in detail. As has been shown earlier, the sedimentation of oligosaccharidases with the microsomal fraction on centrifugation may be a result of destruction of the brush border (Miller and Crane, 1961a, 1961b).

Histochemical data showing that the oligosaccharidase activity is bound to intracellular granules may indicate the site of synthesis and not that of the activity of the enzymes. Finally, the ratio of the concentrations of the reaction products of hydrolysis in the incubation medium and the intracellular fluid is not a reliable criterion for the differentiation of membrane and intracellular digestion.

The third group, data favoring the assumption that hydrolysis is completed on the outer surface of the membranes, is also not free of certain questions. Indeed, it is difficult to exclude the possibility that during the destruction of intestinal cells at least a portion of the active sites may be blocked by some substances present in the homogenate. The resolving possibilities of the electron microscopy–histochemical technique, perhaps, also do not permit unequivocal deductions. However, the observation regarding the effects of convection currents on the hydrolysis of oligosaccharides and peptides and the determination of pore radius of the membranes of intestinal cells did not seem to raise any objections, and this allows the conclusion that hydrolysis occurs on the outer surface of the membrane. The acceptance of this hypothesis permits without any difficulty the use of the idea that the passage of substances across the membrane itself takes place actively.

In the past few years a series of investigations have been performed in our laboratory, the basic problem of which was to discover new methods of differentiating between membrane and intracellular types of hydrolysis and to attempt to determine which of the two in reality takes place. In addition, in certain experiments a verification of the data obtained by the adherents to the intracellular theory has been made in an attempt to understand the reasons for the contradictions which arose in the literature.

Among the first investigations in this series were the studies by DeLaey in Ghent in 1963 and by Iezuitova et al. (1963) in Leningrad. These authors have shown that the phenomena which are well known for heterogeneous catalysis on nonbiological surfaces are characteristic of membrane digestion.

Indeed, like any catalytic reaction on the surface, the hydrolysis on membranes is limited by the slowest of the three stages involved in the reaction: the transfer of the substance from the inner phase to the surface, surface catalytic process, and the removal of the substance from the region of the reaction to the inner phase and border. As a rule, the limiting stage is the rate of transfer of the initial substances from the inner phase to the surface. If so, then mixing should be accompanied by a substantial increase in the rate of membrane digestion.

This was precisely the point demonstrated by the above authors in the case of hydrolysis of starch. Later on it was found that the sharp increase in the rate of digestion during the movement of the fluid in relation to the surface of the mucosa takes place not only for starch but also for sucrose and certain peptides (Iezuitova et al., 1963; Iezuitova, 1963; Timofeeva, 1963; Ugolev et al. 1964a) (Fig. 11A).

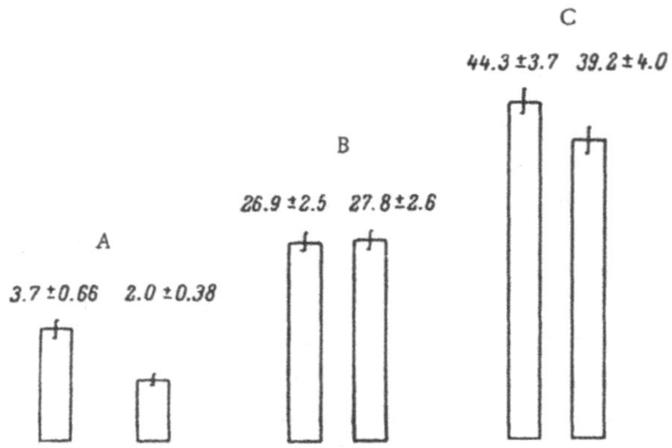

Fig. 11. Effect of mixing on the hydrolysis of sucrose (A) and on passive transport of glucose (B) and galactose (C) in the tissue of small intestine. Numbers indicate the increase in formed hexoses (A) and of hexoses in tissues (B, C) (in mg % per minute; average value ± standard error). In each pair of columns the left bar indicates the effect with mixing and the right bar that without mixing.

Indeed, it was shown that mixing does not affect significantly the rate of penetration of glucose and galactose inside the intestinal cells (Fig. 11B, C). In this connection the experiments of Kirse are of considerable interest; she studied the effect of mixing on the rate of hydrolysis of various oligosaccharides and starch and the rate of transfer of resulting hexoses inside the inverted sac prepared from the small intestine of white rats. She found, in accord with previous data, that the rate of hydrolysis during mixing sharply increases, while the accumulation of the products of hydrolysis in the serous fluid does not increase but, on the contrary, decreases. This phenomenon cannot be explained from the standpoint of intracellular digestion, and it can be understood only on the assumption that the final stages of hydrolysis take place on the surface. In this case the convection currents of the fluid not only bring to the surface enzymes additional amounts of substrates but they also remove a portion of the formed products.

This group of investigations does not exclude the presence of intracellular digestion, but demonstrates that the role of this mechanism is not in any way substantial.

An entirely different principle was employed in the experiments of Nadirova et al. (1965b). For the solution of the question of the localization of the final stages of hydrolysis the following essential scheme of the experiment appeared appropriate: if one blocks the outer surface of the intestinal cells, then the contact of substrates with the

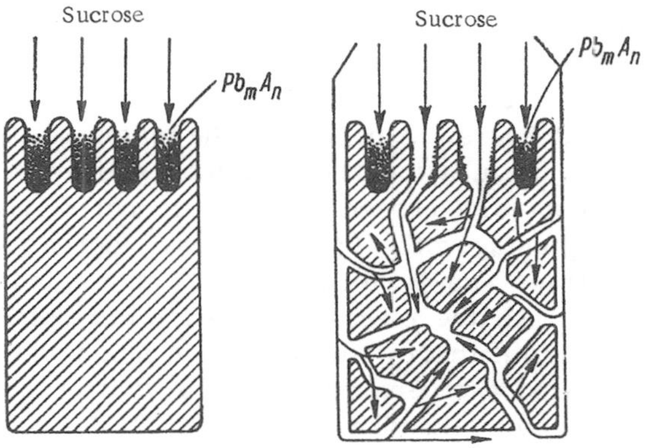

Fig. 12. Essential scheme of experiment demonstrating the conditions of contact of substrates with enzymes of intact (left) and disrupted (right) intestinal cells treated with an isotonic solution of Pb $(NO_3)_2$.

enzymes which are located on the outer surface of the brush border or inside the cell will become impossible (Fig. 12, left). In this manner not only membrane digestion but also intracellular digestion will be excluded. It will be possible to differentiate the localization of the enzymes after the disruption of the cell structure whose brush border has been blocked. Indeed, under these conditions the contact of substrates with enzymes of the outer surface will still be impossible as heretofore, while the activity of the intracellular enzymes of the disrupted cells will be revealed practically in full (Fig. 12, right).

Such experiments were performed and the results are discussed below. The localization of invertase in the intestinal epithelium of white rats was the subject of the experiments. The blocking of the brush border was accomplished by treatment of the surface of the mucosa of the small intestine with a 1% solution of $Pb(NO_3)_2$. The lead ions, like other multivalent ions of other heavy metals, penetrate the cell membranes extremely slowly. No less important is the fact that between the microvilli of the brush border is a precipitate of lead salts (carbonates, phosphates, basic chlorides, and others). The formation of the conglomerates of insoluble salts of lead took place as a result of the diffusion of lead ions into the pores of the brush border from the lead nitrate solution (in which the intestine was preincubated) and of a simultaneous diffusion of the ions of phosphoric, carbonic, and other ions from the intestinal cells to the outside. The localization of the conglomerates of lead salts* was checked by electron microscopy.

*General formula Pb_mA_n.

For the electron microscopic studies pieces of the small intestine after preincubation in a 1% solution of $Pb(NO_3)_2$ or in 0.4% solution of $NaNO_3$ were fixed in a .2% solution of OsO_4 in veronal-acetate buffer (pH 7.4) for 1-2 hours. A portion of the material was fixed in liquid oxygen, ground up in a mortar, and then stained in a 2% solution of OsO_4. The object was passed through alcohol solutions whose concentrations, beginning with 40% to absolute, increased at intervals of 10%. The contrast was performed in absolute ethanol with phosphotungstic acid. Then the pieces of intestine were placed in a mixture of methyl and butylmethacrylates (the ratio of methyl to butyl was 6:1). The polymerization was performed in a thermostat at 55-60°C. Extrathin slices obtained on the ultramicrotome UMT-2 were then examined in electron microscope JEM-5g at 80 kW.

The activity of invertase obtained in duplicate experiments, using a 1% solution of $Pb(NO_3)_2$ (experimental) and in isotonic sdution of $NaNO_3$ (controls), of inverted pieces of the small intestine were compared. This allowed the establishment of the extent of blocking of the surface of the brush border. In the same experiments invertase activity of the homogenates of the experimental and control slices of the small intestine were also compared.

The work was performed on white rats of Wistar strain weighing 150-200 grams. After sacrifice of the animals a section of the small intestine 10-20 cm long next to the duodenum was rapidly removed and washed with a 0.4% solution of $NaNO_3$. The inverted sections were placed on glass rods and immersed in a 1% solution of $Pb(NO_3)_2$ or in a solution of 0.4% $NaNO_3$ (control) isotonically equal to $Pb(NO_3)_2$ for preincubation for 10 minutes in a shaking apparatus. After washing (twice, for 3 minutes each time, in a 0.4% solution of $NaNO_3$), the sections were divided into equal portions of 2 cm each. A portion of the experimental and control sections was ground up in a mortar with liquid oxygen.

The homogenates* and slices of the intestine were incubated in a Ringer solution containing 1% sucrose for 10 minutes at 37-38°C in a Warburg apparatus with constant shaking. The enzymatic activity after incubation was inhibited by rapid cooling.

The activity of invertase was determined by the increase in reducing sugars (method of Nelson as modified by Ugolev).

An additional series of experiments was performed in which sections of the small intestine were ground up in a porcelain mortar with glass. In the experimental test the homogenates were washed out of the mortar with 0.2 ml of a 1% solution of $Pb(NO_3)_2$ together with 2.8 ml of Ringer solution, and in the control test with 0.2 ml of a 0.4% solution of $NaNO_3$ and 2.8 ml of Ringer solution. To each test solution 3 ml of a 1% solution of sucrose was added, after which the basic incubation was performed as described above.

As shown in Fig. 13, the hydrolysis of sucrose by the sections of the small intestine treated with the 1% solution of $Pb(NO_3)_2$ was approxi-

*In the future, the term "homogenate" refers to the obtained detritus which, under electron microscopy, consists of small fragments of cell of irregular shape.

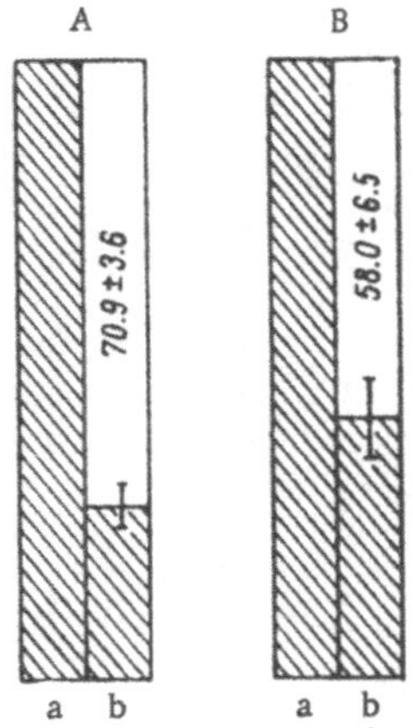

Fig. 13. The decrease in the activity of invertase (numbers inside the column) of intact (A) and homogenized (B) sections of the small intestine after treatment with the isotonic solution of lead nitrate (b) compared to control (a), whose activity is taken as 100%.

mately 1/3 of that in the untreated sections.* The significant decrease in the invertase activity obtained in these experiments could have been related either to the blocking of the pores of the brush border of the intestinal epithelium or to the inhibition by lead ions of the enzyme.

Electron microscopy revealed that the conglomerates of the lead salts were distributed between the microvilli of the brush border of the intestinal epithelium, occasionally tightly surrounding them at different heights (Fig. 14B). The lead penetrates down to the base of the microvilli, forming "plugs." In the cross section it can be seen that lead salts form a granular structure around the microvilli (Fig. 14C). In control experiments no granules of any kind can be seen between the microvilli (Fig. 14A).

Of particular importance is the circumstance that the lead conglomerates were always localized extracellularly. No lead granules were ever observed in the cytoplasm of the cells of intestinal epithelium or in microvilli. It is probable that this can be attributed to the fact that the multivalent ions poorly penetrate the cell membranes as well as to a peculiar "anionic barrier" which arises as a result of diffusion from the cells of anions which form insoluble salts with the lead ion. A portion of the brush border remained free which makes understandable the preservation of a certain amount of the enzymatic activity. A correlation between the number and the localization of the lead conglomerates, revealed by electron microscopy, and the extent of inhibition of the enzymatic activity could not be demonstrated.

While describing the principle of the method, we pointed out that

*In control of experiments it has been shown that the treatment of the intestine with a 0.4% solution of $NaNO_3$ does not affect its invertase activity.

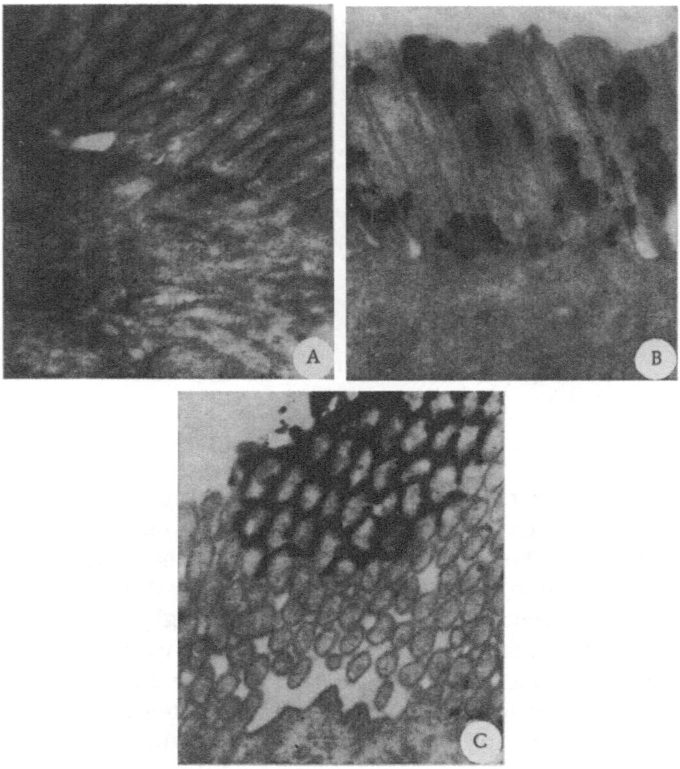

Fig. 14. Localization of conglomerates of lead in intestinal cells. (A) control; section of intestine treated with the isotonic solution of $NaNO_3$; brush border and apical portion of the cell (×30,000). (B, C) experimental; sections of the intestine treated with the isotonic solution of $Pb(NO_3)_2$; longitudinal (B) and cross (C) sections; visible are conglomerates surrounding microvilli at different levels (B:×75,000; C:×38,000) (reduced 15% for reproduction).

the decrease in enzymatic activity as a result of blocking the pores of the brush border does not permit judgment as to the site of the final stages of hydrolysis. To approach the solution of this problem it is necessary to study the effect of disruption of the cell structure on the activity of the enzymes. If the hydrolysis occurs intracellularly then the activity of the treated and untreated homogenates should be approximately the same. On the contrary, in the case of membrane hydrolysis the decrease in the enzymatic activity of homogenates previously treated with the isotonic solution of $Pb(NO_3)_2$, as compared to untreated homogenates, should be approximately the same as the decrease in the enzymatic activity of the treated sections of the intestine compared to that of the untreated sections. As can be seen from Fig. 13, the results of the experiments conform to the second possibility (membrane hydrolysis).

Some decrease in the inhibitory effect (statistically insignificant in our experiments) could be attributed to the fact that in the course of grinding the tissue a portion of the lead is removed from the pores of the brush border. This is confirmed by the data obtained by electron microscopy.

The previously expressed possibility that the lead ions which are localized in the pores of the brush inhibit intracellular invertase has been tested experimentally and proved to be groundless. In a special series of experiments, lead precipitated in Ringer solution in double the concentration generally fixed in the brush border was added to the homogenates of intestinal cells. The activity of invertase under these conditions was equal to 49.2 ± 4.4 mg %, while the control activity was 44.1 ± 4.05 mg %. The difference between these values was not statistically significant.

In analogous experiments for the determination of the effect of Ag^{+} and Cu^{++} on invertase activity it was found that these cations, which practically do not penetrate the cell membranes, in a short time almost completely inhibited the invertase activity.

Thus, the data obtained indicate the validity of the membrane digestion mechanism of the final stages of hydrolysis of sucrose and refute the inductive assumption that this process occurs intracellularly.

The most significant arguments in favor of the intracellular localization of the oligosaccharidase activity were the observations of Miller and Crane (1961a, 1963), who studied the distribution of hexoses between cells and the incubation medium in the course of hydrolysis of sucrose by intestinal epithelium in vitro. If one excludes active transport, then in the case of extracellular hydrolysis the concentration of the formed hexoses in the incubation medium should be higher than in the intracellular fluid. During intracellular hydrolysis the concentration gradient should be directed from the cell to the surrounding medium. Miller and Crane (1961a, 1963) observed a higher concentration of hexoses in the tissues than in the surrounding medium (Fig. 15). However, as we will see later, the data of Miller and Crane did not prove the intracellular localization of the hydrolysis of oligosaccharides. By virtue of a whole series of circumstances the authors employed methodological conditions under which membrane digestion "simulates" the distribution of hexoses which, according to Rothstein, is characteristic of intracellular digestion.

Later we will discuss the theoretical considerations and experiments with an electric model of the cell which indicated that the concentration

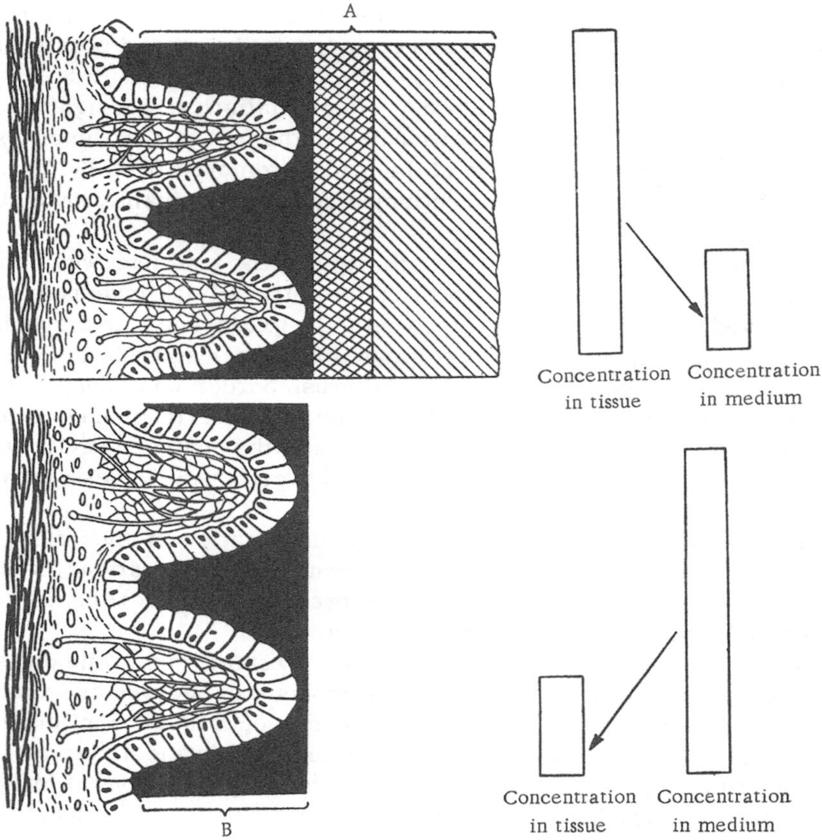

Fig. 15. The scheme of experiments of Miller and Crane (A, average values) and of Iezuitova, De Laey, and Ugolev (B, membrane test).

criterion as it has been used is not reliable for the differentiation of membrane and intracellular digestion.

For the description of the localization of the action of oligosaccharidases, the data on the ratios of the concentrations in the immediate vicinity to the membrane of the intestinal cells are of the greatest significance because the concentration of the products of the reaction will be diminished in proportion to the distance from the surface of the intestine (Fig. 15). This is particularly important since the presence of villi and microvilli slows down the diffusion of the products of the reaction into the basic volume of the incubation medium and enhances the penetration of the substances inside the cells.

In our laboratory (Ugolev et al., 1964a) the data of Miller and Crane were reproduced under their conditions in which the concentration of substances in the whole volume of the fluid was compared with that inside the cells. Furthermore, a method was developed which permitted the determination of the concentration of hexoses formed on hydrolysis of sucrose immediately next to the membrane.

The experiments were performed on strips of small intestine of white rats of Wistar strain. The strips were shaken for a short time in a Ringer solution containing 0.02% NaF (to exclude active transport) and 2% sucrose. The strips were then removed from the solution under conditions which prevented dehydration and incubated for 2-15 minutes at a room temperature of 18-22°C. At the end of incubation the fluid was removed by filter paper which covered the mucous membrane and the concentration of hexoses in the tissue and in the fluid collected on the filter paper was determined.

Experimental Technique. The intestine was rapidly removed, washed, and then placed in a cold Ringer solution containing 0.02% NaF to exclude active transport. The 3- to 4-cm-long and 0.7- to 1.4-cm-wide pieces of intestine were washed and equilibrated with a 2% solution of sucrose for 30 seconds. After the removal of the intestine from the sucrose solution, the excess fluid was allowed to drain off, and the intestine was placed in a nylon sack to prevent dehydration. After incubation, the sack was opened and a filter paper was placed on the surface of the mucosa; the other side of the intestine was still covered by the nylon film. Within a short time of contact with the mucosa, a certain amount of fluid was thus extracted; this amount was determined by the difference in weight of the filter paper before and after the soaking.* Pieces of the intestine were cooled, weighed, and homogenized, and the hexose content in the homogenate was determined by the method of Nelson (1944).

In special experiments it has been shown that the determination of hexose in the fluid surrounding the intestine by generally accepted methods and by the use of filter paper gave similar results.

In the first portion of the work the conditions employed by Miller and Crane were reproduced, i.e., the incubation of the pieces of intestine was made in volumes of sucrose solutions several times the weight of the intestine (Miller and Crane, 1961a, 1963). In this case the data of Miller and Crane (shown in Fig. 16) were confirmed. In addition, in accord with these authors, we observed that the concentration of the hexoses formed in the tissue was independent of the volume of the incubation medium, while the concentration of hexose in the latter was inversely proportional to its volume.

The objective of the second portion of the work was to establish the concentration of monosaccharides in the immediate vicinity of the outer surface of the membrane. Certainly, if the hydrolysis is performed intracellularly then in this case the concentration of hexoses in the outer fluid should be smaller than that in the intracellular fluid (Fig. 15A). Besides, with the increase of intervals of incubation, there should be a relative increase in the concentration of hexoses in the extracellular fluid, as has been demonstrated by Miller and Crane under the conditions of their experiments. However, as can be seen

*Employing histological control it was shown that in these experiments the structure of the villi is not damaged.

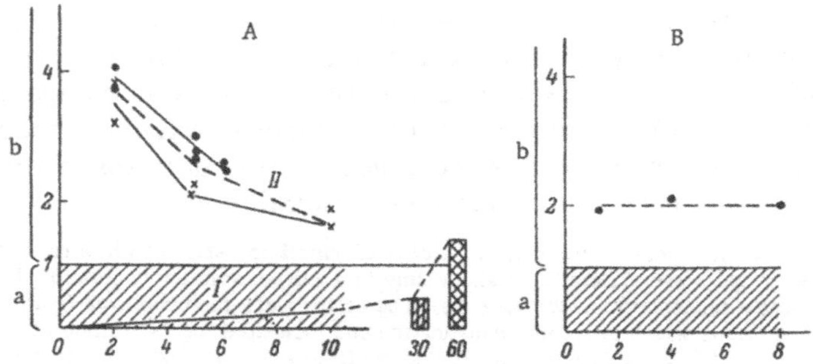

Fig. 16. Relationship of concentrations of hexoses, formed during hydrolysis of sucrose, in the incubation medium and in the intracellular fluid (A) and hexoses, formed during hydrolysis of sucrose, by the inverted piece of intestine (B). In A—abscissa: time of incubation (in minutes); ordinate: ratio of the concentration of hexoses in the incubation medium to the concentration of hexoses in the intracellular fluid; I) during the determination of hexoses in the large volume of sucrose solution (according to Miller and Crane, 1961a); II) during determination of hexoses in the immediate vicinity of the surface of the cells. B—abscissa: length of intestine slice (in cm); ordinate: same as in A (a) zone typical of intracellular digestion; (b) zone typical of extracellular digestion.

from Fig. 16, the concentration of monosaccharides in the surrounding medium in contact with the surface of intestinal cells is higher than in the intracellular fluid. This by itself contradicts the assumption that the hydrolysis of sucrose occurs intracellularly, and it provides evidence in favor of membrane digestion.

Even more important is the circumstance that the ratio of the concentrations of hexoses in the extracellular fluid to that in the intracellular fluid, on continuation of incubation, definitely decreases. This result is possible only if the hexoses are initially formed outside the cells or on their outer surface, and only then penetrate the cell. As has been noted above, in the case of intracellular hydrolysis this ratio increases with the increase of time of incubation.

The basic results obtained by Miller and Crane were reproduced by incubation of comparatively small pieces of the intestine in large volumes of solution (Miller and Crane, 1961a, 1963). These data were interpreted by the above-mentioned authors as proof of the intracellular hydrolysis of disaccharides. However, the determination of the concentration of hexoses in the immediate vicinity of the intestinal cells performed by special methods and the comparison of the results with the concentration of hexoses in the intracellular fluid has shown that a gradient exists which is the reverse of that described by Miller and Crane. In reality the concentration of substances in the fluid which is located in the immediate vicinity of the outer surface of the cells is

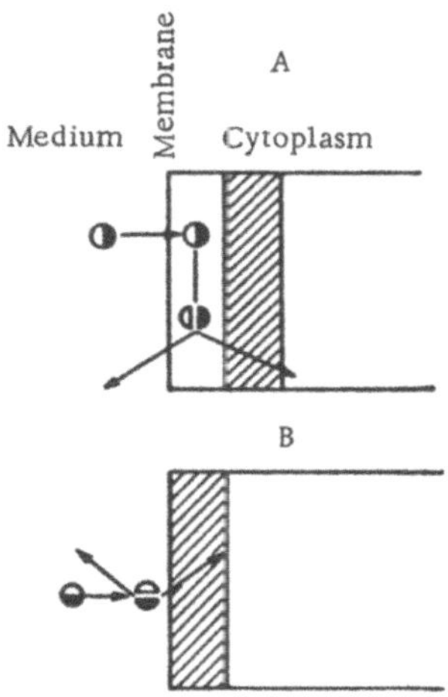

Fig. 17. Localization of the hydrolysis of sucrose. (A) according to Crane (1962) and (B) on the basis of our experiments.

higher than in the intracellular fluid, which is possible only if the hydrolysis of sucrose occurs on the outer and not on the inner surface of the intestinal cells (Fig. 17).

The dynamics of the distribution of concentrations in time even further confirm this point of view. In the evaluation of the method employed by us, we should mention that one cannot claim that it reveals the concentration of hexoses in the fluid present in the pores of the brush border; our data, therefore, are only approximations of the characteristics of a true gradient of concentrations. It is logical to assume that the concentration of hexoses in the immediate vicinity of the outer surface of cell membranes must be even higher than the one we observed. The results obtained by Miller and Crane reflect only dilution rather than true distribution of substances on both sides of the cell membrane.

Indeed, in the experiments by Meboldt and Ugolev (cited by Ugolev, 1964) employing an electric model of an intestinal cell, it has been shown that, depending on the ratios of volumes of intracellular and extracellular fluids, membrane digestion may give reflections typical of intracellular and extracellular hydrolysis (Figs. 18 and 19). These model experiments decisively prove that the results of Miller and Crane are not proof for the intracellular localization of the final stages of hydrolysis of carbohydrates but reflect regularities which are typical for membrane digestion under rigidly controlled conditions of the experiment.

The results obtained in our experiments are in good accord with the observations of Miller and Crane (1963) which, however, were not discussed by these authors. They have shown that during a very prolonged incubation the concentration of hexoses formed in the surround-

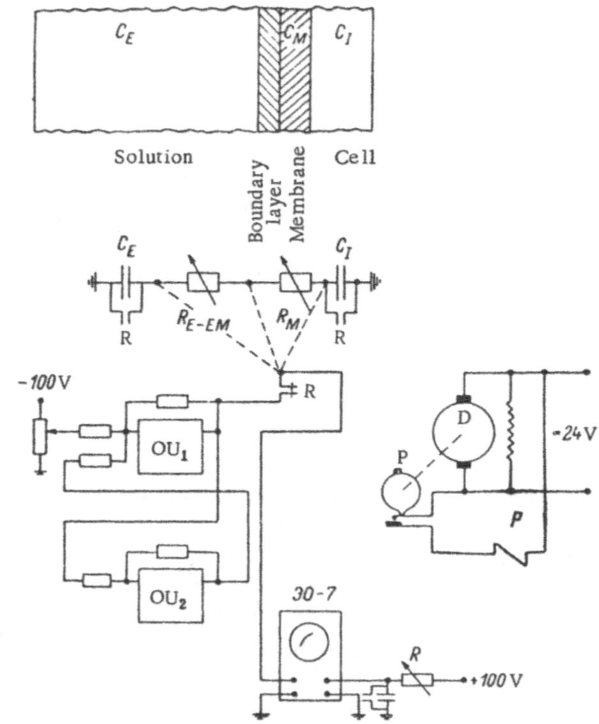

Fig. 18. The principal and electric scheme of a model of an intestinal cell. Top: C_E) concentration of product in solution; C_I) concentration of product inside the cell; C_M) concentration of product on the outer surface of the membrane. Bottom: D) MN-250 motor; P) interrupter; C_E) KBG condenser; C_I) KBG condenser; R) RC-52 relay; R_{E-EM}) resistance (0-40 kilohms); R_M) resistance (0-40 kilohms); and OU_1, OU_2) operational accelerators.

ing medium may be higher than that formed in the intracellular fluid. This is impossible if the digestion took place intracellularly, but understandable if membrane hydrolysis was taking place.

In the experiments of Miller and Crane, glucoseoxidase, which was also employed as proof of the intracellular hydrolysis of disaccharides, retained its significance up to the present. The essence of the experiment consisted in the observation that glucoseoxidase added to the incubation solution does not penetrate intestinal cells and for this reason cannot cleave glucose formed from the disaccharides if this takes place intracellularly. However, glucoseoxidase will oxidize glucose if the formation of the latter takes place on the surface of the membranes.

The authors found that their results are in accord with intracellular hydrolysis. However, the authors were using unpurified preparations of glucoseoxidase and the enzyme could not penetrate the pores of the brush border.

It is clear that a thorough study of the hydrolysis of di-, tri-, and tetraglucosides by the preparations of the small intestine by Ruttloff et al. (1965) employing pure preparations of glucoseoxidase with a molecular weight of about 150,000 is of specific significance. It was found that the glucoseoxidase prevents the glucose, formed during hydrolysis, from entering inside the cells. Thus, these experiments provide important proof for the fact that the transformation of oligomers into monomers occurs at the surface of cell membranes and not intracellularly. Thus these experiments not only provide additional serious support for the membranal nature of the final stage of digestion, but no less importantly explain the contradictions which had existed in the literature.

As has been mentioned above, the experiments with mixing showed that the hydrolysis of dipeptides occurs at the outer surface of cell membranes. However, such an important question as the localization of the final stages of hydrolysis of proteins made it appear imperative to investigate by means of other methods which embodied definite and unambiguous criteria for the intracellular as well as the membrane hydrolysis.

For the solution of this problem we developed an approach, novel in principle (Kushak and Ugolev, 1966), based on the analysis of the relationships of the products of hydrolysis inside intestinal cells. These relationships will be altered depending upon where the hydrolysis of dipeptide takes place—from the outer or the inner side of the cell membrane; in the absence of active transport, one may consider it a simple diffusion membrane.

The effect of the size of the water-soluble molecules on the rate of their diffusion across cell membranes is a well-known fact. It has been established, in fact, that the smaller their molecular volume, the faster molecules of amino acids penetrate the cell membranes. Thus, for instance, according to data of several authors, glycine penetrates the intestinal cell faster than leucine. The analysis of the relationships of the products of hydrolysis of dipeptides consisting of amino acids of various molecular weight presents a real possibility to differentiate the hydrolysis inside the cell from that on the outer surface of intestinal membranes.

Figure 20, I and II, presents the theoretical assumptions of the experiments. If the hydrolysis of dipeptides takes place intracellularly, then the dipeptides must penetrate the cells where the amino acids are

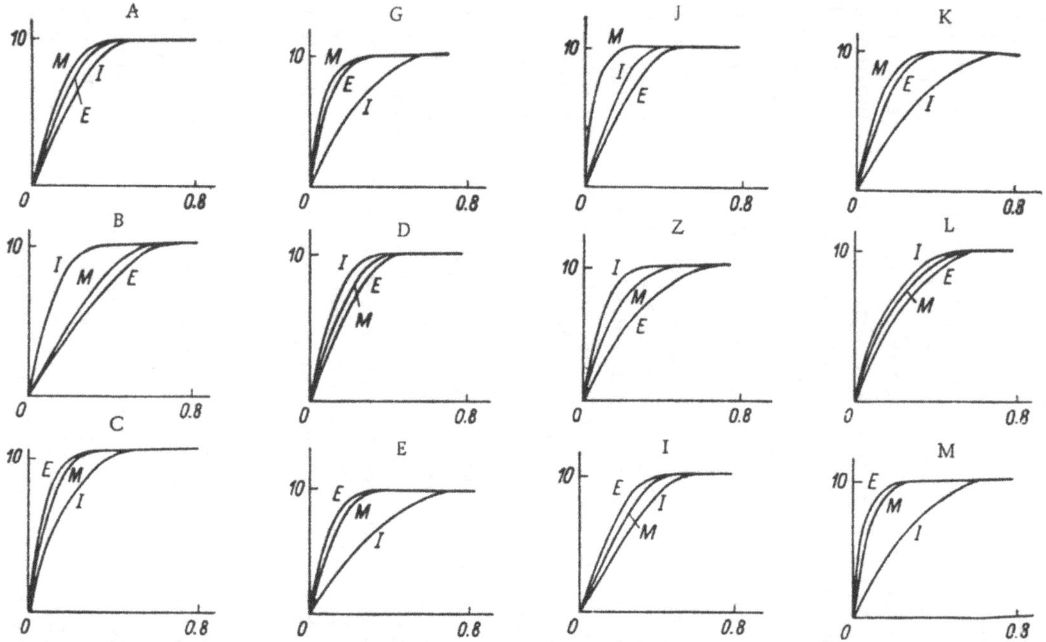

Fig. 19. Dynamics and the ratio of concentrations of products of the reaction at various points of the electric model: inside the cell (I), in the surrounding medium (E), and on the outer surface of the membrane (M). Abscissa: time of the reaction (in seconds); ordinate: concentration (tension in volts). The formation of products of the reaction in membrane layer (A, G, J, K), inside the cell (B, D, Z, L), and in solution (C, E, I, M). Time of load = 0.12 seconds; time of charge = 1.6 seconds; reamer, 12 cm. On the curves A–E, R_{E-EM} = 4 kilohms, R_M = 40 kilohms; on the curves A, B, C, C_E = 8 μF, C_I = 1 μF; on the curves G, D, E, C_E = 1 μF, C_I = 8 μF; on the curves J–M, R_E–EM = 20 kilohms, R_M = 40 kilohms; on the curves J, Z, I, C_E = 8 μF, C_I = 1 μF; on the curves K, L, M, C_E = 1 μF, C_I = 8 μF.

formed in equimolar amounts. Since the rate of exit of "small" amino acids is higher than that of "large" amino acids, the intracellular fluid will contain higher concentrations of the "large" amino acids. If the hydrolysis of a dipeptide takes place on the outer surface of the cell membrane, then the amino acids will diffuse across the latter independently of each other at rates inversely proportional to their molecular volume. In this case the concentration of "small" amino acids in the intracellular fluid, on the contrary, will be higher than that of the "large" amino acids.

These data were confirmed by us in special experiments dealing with the distribution of amino acids formed during the hydrolysis of dipeptides in media which were separated by cellophane membranes as well as in studies of the entry of amino acids inside intestinal cells from mixtures equivalent to the mentioned peptides.

The experiments were performed on inverted sections of the small intestine of white rats (Wistar strain), 3-cm long, treated with a 0.02% solution of NaF to inhibit active transport. The sections were incubated in 4 ml of 0.01 M Ringer salt solution of certain dipeptides consisting of amino acids of various molecular weights (glycyl-L-leucine, glycyl-L-tyrosine, glycyl-DL-α-alanine) as well as of equimolar mixtures of amino acids comprising these dipeptides. The incubation was conducted for 0.5 to 15 minutes, with mixing, in a Warburg apparatus (temperature 38°C). After the incubation the slices were rapidly cooled and amino acids were isolated from the mucosa by the usual methods. The amino acids were separated by thin-layer chromatography on silica gel-gypsum using phenol-water (75:25) for glycyl-L-leucine and glycyl-L-tyrosine and propanol-25% ammonia (70:30) for glycyl-DL-α-alanine.

Quantitative analysis was performed by means of a recording densitometer (Jouce, Loeb and Co., Ltd.). The increase in amino acid concentration was determined by comparison with slices alone which were simultaneously incubated in the Ringer solution.

As can be seen from Fig. 21, during the hydrolysis of glycyl-DL-α-alanine, glycine and alanine accumulated in the intracellular fluid at about the same rate, which could be expected since the amino acids have approximately similar molecular weights.

During the incubation of solutions of dipeptides consisting of amino acids of different molecular weights (glycyl-L-leucine and glycyl-L-tyrosine) (Figs. 20 and 22), glycine accumulated faster in the intracellular fluid than either leucine or tyrosine which, in accord with theoretical considerations, is characteristic of membrane hydrolysis.

It turned out that the increase in the concentration of individual amino acids in the intracellular fluid during incubation of solutions of glycyl-L-leucine and glycyl-L-tyrosine occurs similarly to that during the incubation of equimolar mixtures of the component amino acids (Fig. 22).

In describing the principle underlying this work, we indicated that such kinetics of accumulation of "large" and "small" amino acids are possible only if their formation takes place outside the cell membrane, and not inside the cell. In the latter case the accumulation of the "large" amino acids should have taken place faster than that of

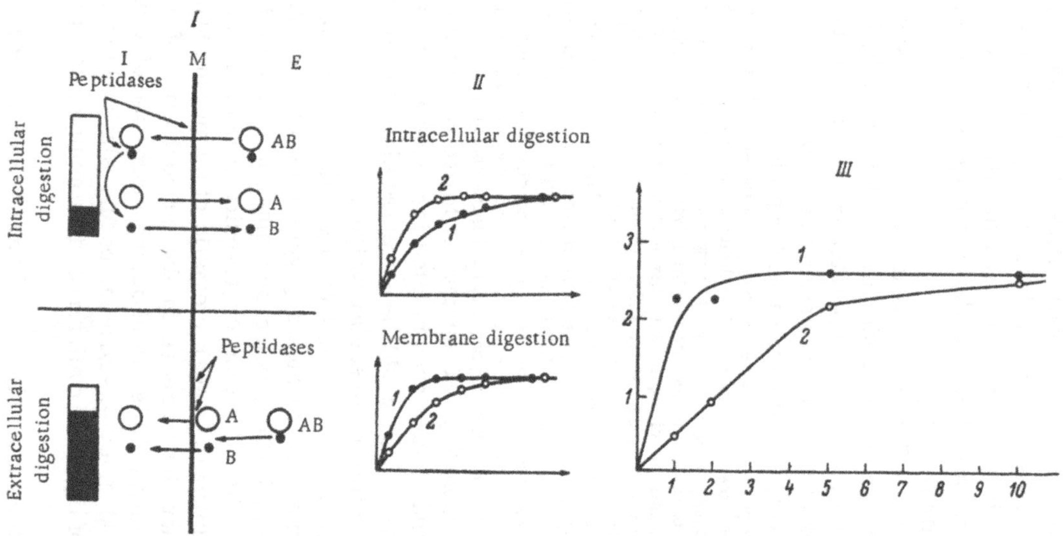

Fig. 20. Relationship of "large" and "small" amino acids, formed during the extra- and intracellular hydrolysis, in intracellular fluid. (I)—AB) dipeptide; A) "large" amino acid; B) "small" amino acid; I) intracellular fluid; M) membrane; E) extracellular fluid. (II, III)—Kinetics of accumulation of glycine (1) and leucine (2) in tissues of the small intestine during the hydrolysis of glycylleucine: (II) theoretical variant, (III) experimental results. Abscissa: Time of incubation (in minutes); ordinate: increase in amino acids in the sample (in micromoles).

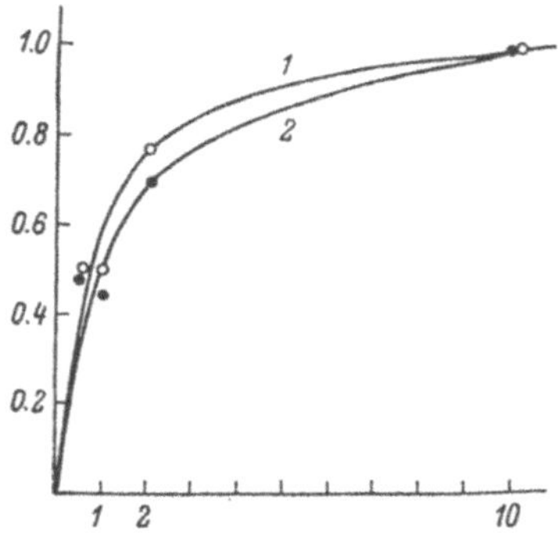

Fig. 21. Kinetics of accumulation of amino acids in tissues of the small intestine during hydrolysis of glycyl-DL-α-alanine. Abscissa: time of incubation (in minutes); ordinate: content of amino acids (in relative units in which a unit is taken as the content during the stationary phase of the experiment). (1) Glycine; (2) alanine.

the "small" amino acids. A confirmation of this is the fact that the concentration of glycine in the incubation mixture during the first minutes of incubation is small, whereas that of leucine is already present in significant amounts. In addition, in spite of the high content of the dipeptide in the incubation mixture, in none of the experiments could intact glycyl-L-leucine be detected inside the intestinal cells.

The data obtained contradict the assumption regarding the intracellular localization of the hydrolysis of dipeptides and lead to the conclusion that this process is accomplished at the outer surface of the brush border.

We now summarize the problems discussed in this chapter.

An ample basis now exists for the assumption that membrane digestion plays an extremely important role in the hydrolysis of nutrients in the intestines of higher animals and man. The hydrolysis on the surface of the intestinal epithelium is induced by enzymes adsorbed from chyme, the enzymes which are mainly secreted by the pancreas (amylase, lipase, proteases), and by truly intestinal enzymes (peptidases, oligosaccharidases, alkaline phosphatases, and others).

The circumstance that the activity of enzymes adsorbed from chyme takes place on the outer surface of the small intestine, and not intracellularly, will not, perhaps, raise any questions. The very fact of the reversibility of adsorption and the disappearance of "membranal" effects during prolonged perfusion is sufficient proof of this. Also, the molecules of enzymes and the molecules of substrates such as proteins and polysaccharides do not penetrate the cell membranes in any significant quantities.

In the investigations cited in this chapter (other investigations are considered in Chapter IV), it was shown that the role of membrane digestion in the cleavage of proteins and polysaccharides is not smaller than that played by the classic cavital hydrolysis. Later on we show

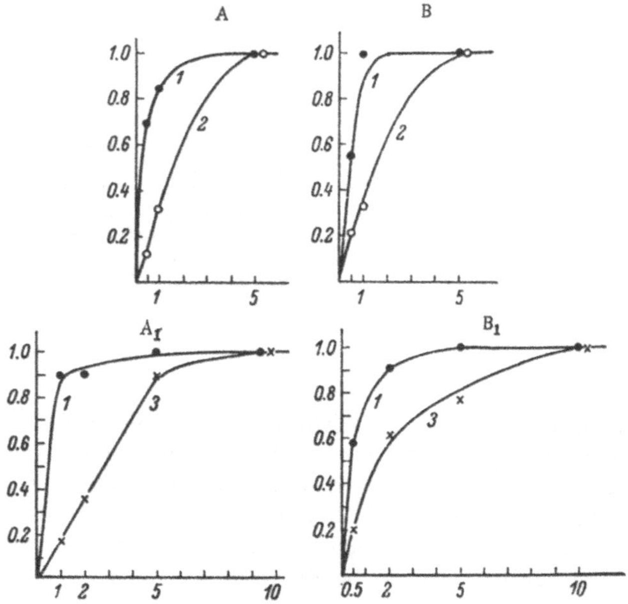

Fig. 22. Kinetics of the accumulation of amino acids in tissues of the small intestine during incubation of solutions of glycyl-L-tyrosine (A), glycyl-L-leucine (A_1), glycine and tyrosine (B) and glycine and leucine (B_1). (1) Glycine; (2) tyrosine; (3) leucine. The rest of the notation is the same as in Fig. 21.

that the significance of membrane digestion is greater the smaller the molecule of the substrate.

Considerably more difficult for experimental analysis were the attempts to determine the site of the final stages of hydrolysis. Possibly these difficulties appeared particularly great because the ideas regarding intracellular hydrolysis were used in cases for which there were no direct data. However, even for oligomers several methods were found which permitted differentiation of membrane from intracellular digestion. When this was done it was found that hydrolysis of oligomers takes place primarily or completely on the outer surface of the intestinal cells.

Such localization of the intermediary and final stages of digestion, as will be shown later, creates extremely favorable conditions for the transition from the processes of hydrolysis to the processes of active transport across the cell membranes.

Thus, all the methods employed which permit the establishment of the localization of the final stages of hydrolysis lead to the conclusion that this process is accomplished before the oligomers penetrate the cell membrane. Let us recall the entire system of the argument.

The hydrolytic enzymes are concentrated in the region of the brush border. However, the histochemical data do not permit deciding the question because of the poor resolving possibilities of light microscopy. The coupling of histochemistry with electron microscopy furnishes ground for the belief that the alkaline phosphatase is more likely to be localized on the outer rather than the inner surface of the cell membrane which borders the microvilli. However, the insufficiency of the resolution and no less the diffusion of the contrast material make it difficult to answer this question more exactly.

The isolation of the brush border is not by itself novel with respect to enzyme topography as compared to histochemistry and it only permits showing that the bulk of the enzymes is concentrated in the region of the intestinal surface.

All other facts indicate that the final stages of hydrolysis take place not inside the cell but on its surface. Indeed, the precise gradient of the products of hydrolysis is directed from the outer to the inner side of the cell. An increase in hydrolysis with increase in convection testifies that the limiting process is the diffusion from the depth of the phase to the surface of the cell membranes, and not the passage across the membrane; that is, in other words, the hydrolysis of disaccharides and dipeptides takes place on the outer surface and not inside the cell.

Blocking of the outer surface of the brush border and subsequent determination of the enzymatic activity also lead to the conclusion that hydrolysis takes place on the membrane.

Study of the effective radius of the pores leads to the conclusion that the penetration of sucrose inside the cells is not possible or at any rate takes place extremely slowly. If so, then the high rates of hydrolysis in vivo can be explained only by the assumption that this mechanism does not require the passage of disaccharides across the cell membranes.

If the final stages of hydrolysis occur intracellularly, then in the cytoplasm large monomers must accumulate; if the hydrolysis occurs on the membrane, small monomers should accumulate. As has been shown, the latter is the case.

Increasing the density of the membranes using calcium, and decreasing the permeability toward sodium and sugars under these conditions make one reflect as follows. If hydrolysis takes place after the oligomers pass across the diffusion barrier, then the intestine treated with calcium should display lowered enzymatic activity and, besides, the accumulation of the products of hydrolysis should be greater inside the cell, since reverse diffusion is slowed down. In reality the opposite is true: hydrolysis increases and the amount of sugar entering inside the cell sharply decreases. Thus, we once more must conclude that hydrolysis of oligomers and dimers takes place via membrane digestion.

Biological systems are complex, the assumptions of the investigators are occasionally unreliable, and in the majority of cases they do not exhaust the essence of the phenomenon. For this reason the use of any single method for the solution of this or other biological problems does not give absolute certainty regarding the correctness of the conclusion. All one can say is that the application of a given method makes a selected point of view more or less probable. In this context, verification by several independent methods, based on various approaches, could be quite useful.

It was this that my colleagues and I attempted to accomplish in the past few years, checking by ever newer methods the localization of the final stages of hydrolysis.

Chapter III

Physicochemical and Structural Aspects of Membrane Digestion

Powerful contact hydrolysis on the border of intestinal cells and chyme is possible only because of the extensively developed surface of the cell membranes. And it is clear that the surface provided by the villi (visible to the naked eye) is not sufficient to account for the hydrolysis. For this reason it should be sufficient to conclude on the basis of the physiological experiments that submicroscopic porosity exists which, firstly, sharply increases the active surface and, secondly, imparts to the surface of the intestinal epithelium properties of the pore catalyst.

The existence of such submicroscopic pores was proved by Granger and Baker (1949, 1950) over 15 years ago; they were the first to investigate the epithelium of the small intestine of rats by electron microscopy. What in the light microscope appeared to be cuticle, in reality turned out to be a large number of cell protrusions. According to the data of the above authors, the protrusions have a height of about 0.62 micron and a diameter of about 0.08 micron. The number of these protrusions can be gauged from the following data: on each cell there are about 3000 protrusions (later they were called microvilli), and on each mm^2 of the surface of intestinal epithelium there are about 50 to 200 million of them. It has been calculated that because of microvilli the absorbing surface of the intestine* is sharply increased about 30-fold.

Later on the microvilli were described by many investigators (Dalton et al., 1950, 1951; Schiebler, 1955; Goldin, 1956; Zetterqvist, 1956; Johnston and Wiggans, 1958; Shestopalova, 1959; Barrnett, 1959; Hartmann et al., 1959; Palay and Karlin, 1959a, 1959b; Shestopalova et al., 1960, 1961; Schmidt, 1961; Holt and Miller, 1961; Lesher et al.,

*Workers who were investigating the ultrastructure of the brush border in reality stressed the significance of these particular submicroscopic structures.

1961; Brown, 1962; Gottlieb et al., 1962; Millington and Finean, 1962; Mohiuddin, 1962; and Williams, 1963).

What are microvilli? Microvilli are cylindrical protrusions enclosed in a cell membrane (Fig. 23). Their height is about 7-10 times greater than their diameter (height, 0.97 ± 0.09 micron; diameter, 0.1 ± 0.003 micron). According to Granger and Baker (1949, 1950), Sjöstrand (1956, 1963), and Zetterqvist (1956), microvilli are cytoplasmic structures covered on the outside by a typical cell membrane of about 100 Å thickness and consisting of three layers, outer osmophylic, inner osmophylic, and an intermediary layer which does not adsorb osmium (Figs. 24 and 25). This picture is well in accord with all data regarding the lipoprotein structure of the membranes (Danielli and others).

Approximately similar data regarding the structure of the brush border of the rat small intestine were obtained by Goldin (1956) and Shestopalova (1959) and Shestopalova et al. (1960, 1961). According to Goldin, microvilli are elements 0.5-1 micron in length and 0.06-0.08 micron in width. In contrast to the data of Granger and Baker and

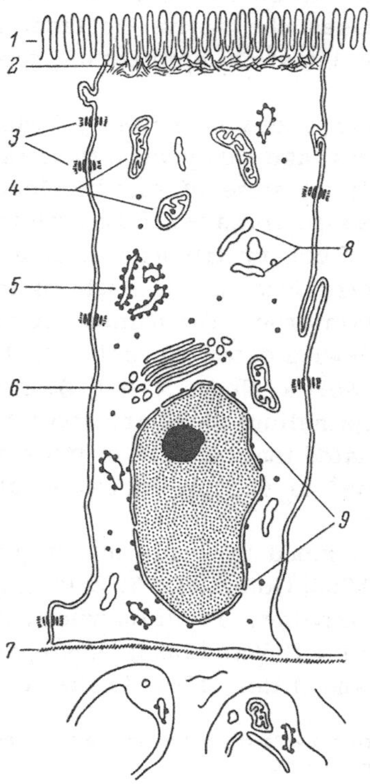

Fig. 23. Schematic representation of the intestinal cell (according to Trier et al., 1963, page 769, Fig. 2). (1) Microvilli; (2) terminal net; (3) desmosomes; (4) mitochondria; (5) granular reticulum; (6) Golgi apparatus; (7) basal membrane; (8) smooth reticulum; (9) nucleus.

other investigators, Goldin and later Shestopalova considered microvilli to be not homogeneous structures but rather submicroscopic hollow cylinders inside which a small canal serves for the absorption. In many cases, as is pointed out by Goldin and Shestopalova, microvilli are not distributed uniformly on the surface of the cells but in groups (small bunches). As regards the functional significance of the microvilli, all the above-mentioned authors consider them to be a special apparatus which ensures intensive absorption.

Certain differences noted by various authors describing the details of the structures depend in part on the imperfection of the method employed and in part on the fact that the investigations were performed on different objects (Table 1) and, finally, probably on the fact that the functional state of the intestinal cells, whose role has not been sufficiently studied, is of considerable importance. Thus, some investigators consider the structure in starving and fed animals to be sharply different; this should be of great interest for the understanding of the intimate processes of digestion and absorption (Palay and Karlin, 1959a, 1959b).

The brush border on the surface of the intestinal cells is a universal structure and is widely distributed in the animal world. Also, the basic features of the structures of microvilli are retained in the most

TABLE 1

Data for the Determination of Sizes, Number, and the Surface of Microvilli (according to Shestopalova et al., 1960, p. 36)

	Sigmodon hispidus	*Macacus rhesus*
Height of microvilli	From 0.55 to 0.93 micron (different sizes)	1.10 microns (different sizes)
Microvilli (transversely)	0.08 micron (800 Å)	0.05 micron (500 Å)
Ratio of height to transverse length	7:1 or 11.5:1	22:1
Lumen between neighboring microvilli	0.02 micron (200 Å)	0.012 microns (120 Å)
Width	–	0.02 microns (200 Å)
Surface of individual microvillus	0.20 micron^2	0.17 micron^2
Number of villi per micron^2 of free cell surface	145	180-216
Surface of microvilli per micron^2 of free cell surface	29 microns^2	30.6-36.7 microns^2

primitive invertebrate and in higher animals, including man. Zetterqvist (1956) mentions the unpublished data of Sperber who described the brush border of rats, chicks, frogs, and certain species of annelids.

Tokin (1959) investigated in detail the ultrastructure of the brush border of *Parascaris equorum*. Most striking was the almost complete identity of the structure of the microvilli of these animals with that of the microvilli of higher vertebrates. As in the mammals, the brush border represents a conglomerate of numerous protrusions of the cytoplasm of the intestinal epithelium. As a result of this a 35-fold increase in the surface is accomplished (a similar increase in the surface is characteristic of the intestine of higher animals). Each protrusion is enclosed by a three-layer membrane which consists of the inner and outer osmophylic layers and the intermediate homogeneous layer.

It appears remarkable that the sizes of the pores of the brush border approach those of the many porous catalysts which are employed in industry, for instance, in the aluminumsilicate industry (Wheeler, 1955). To compare the surfaces of various catalysts one usually calculates the area of the surface per gram weight of the catalyst. Ugolev (1963) calculated the approximate surface per gram of the brush border. The results showed that it is 300,000 to 800,000 cm^2 (the specific surface of the spongy platinum is about 300,000 cm^2 per gram).

The pores of the catalysts play a significant role not only because they increase the active surface, but also because connected with them are the phenomena of transport of substances in the process of catalytic action in those cases in which the sizes of the pores are of about the same order of magnitude as the size of the molecule. In addition, the diameter of the pores determined the size limit of the molecules which penetrate the pores. Thus, the fact that microvilli are on the intestinal surface transforms it into a peculiar porous catalyst.

The very fact of the existence of membrane digestion and the extreme intensity of the hydrolytic processes taking place on the surface of the brush border indicate the great significance of this mechanism. Certain basic functional peculiarities of this mechanism depend on the fact that the cleavage of nutrients takes place on the same surface which also performs the absorption. In this chapter we discuss membrane digestion from the standpoint of reactions taking place on surfaces.

In his classical monograph, Adam (1947) noted the similarity of cell surfaces and surfaces of the solid phase. The course of the reactions taking place on the border of the liquid and solid phases differs markedly from that taking place in the liquid phase. Apparently, it is most appropriate to compare the surface of the intestinal cell, containing the lyophylic as well as lyophobic groups which adsorb various substances, with heterogeneous porous surfaces.

It is also proper to note that the very fact of high catalytic activity of nonliving surfaces often depends on their heterogeneity.

The state in which a substance exists on the border of the phases differs from the state of the same substance in the liquid phase in many respects, particularly in the energy level. As a rule, the surface molecules possess greater energy than those located inside the phase. This permits stating that energy is concentrated on the surfaces (Adam, 1947; Roginskii, 1948; De Boor, 1959; Germain, 1961; and Miner, 1954).

To the biologist, the surfaces of complex solutions are of particular interest. Since the system aims toward reduction of surface energy, a series of processes arises which proceed with a reduction in surface tension. This is revealed in the phenomena of positive and negative adsorption. The majority of inorganic salts cause an increase in the surface tension of water, while organic substances, with few exceptions, reduce surface tension and hence are adsorbed on the phase border. The latter circumstance permits the assumption that very favorable conditions exist for the passage of nutrients from the chyme to the surface of the intestinal cell, that is, from cavital digestion to membrane digestion.

The phenomena of positive and negative adsorption lead to considerable differences in pH on the phase border and in the liquid (Adam, 1947; Roginskii, 1948; Germain, 1961; and others). It is known that the difference in pH in the liquid and on the phase border of biological systems may reach 1.5-2.00 units (McLaren and Babcock, 1959). Thus, the conditions for membrane digestion must differ sharply from those of cavital hydrolysis.

The selective adsorption of various cations and anions of the phase border leads to a significant phase potential (Adam, 1947; Pchelin, 1951; Danielli, 1952, 1954a, 1954b; Hewing, 1956; Sheludko, 1960) which in turn considerably affects the kinetics of chemical reactions and the equilibrium point.

In contrast to molecules which are inside the phase, the molecules on the surface in the majority of cases are in the oriented state to which Langmuir drew attention.

Since the surface of the intestinal cell is heterogeneous, it can be assumed that the orientation of nutrient substances in relation to the surface (and to the enzymes fixed on it) will undergo changes depending on their lyophilic or lyophobic properties and the charge of the given

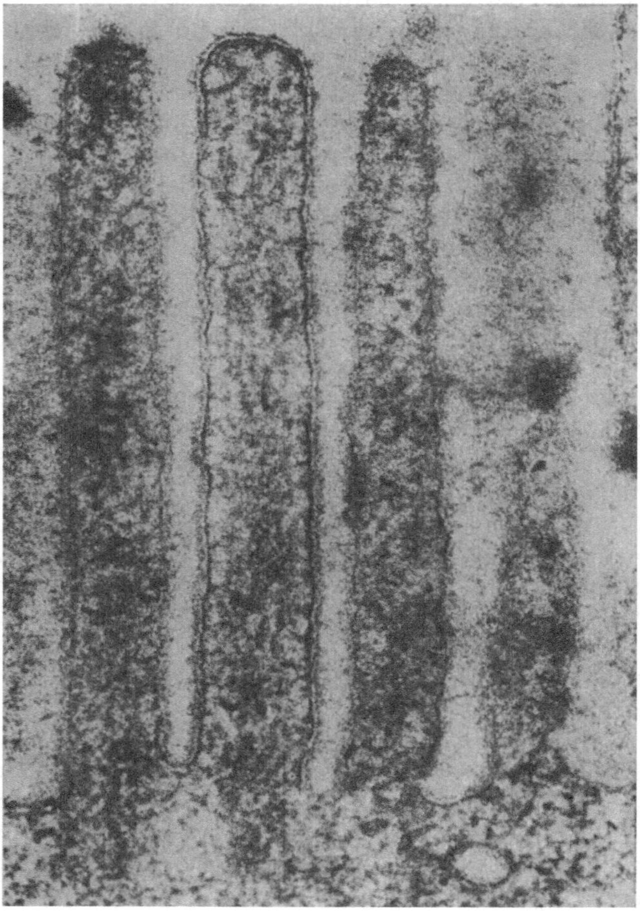

Fig. 24. Microvilli in brush border during absorption of fat. Asymmetric three-layer membrane is plainly seen. × 240,000 (reduced 15% for reproduction) (according to Sjöstrand, 1963, page 530, Fig. 8).

point of the surface. This permits the assumption that favorable conditions arise at the surface for the treatment of the nutrient substances in several aspects. A more detailed treatment of the problem of adsorption of substances on biological structures is given in reviews by Troshin (1956), Nasonov (1959), McLaren and Babcock (1959), McLaren et al. (1963), and by us (Ugolev, 1963). Finally, at the phase border one encounters deformation of molecules.*

* This problem is discussed in detail in various physicochemical reviews. For its significance to hydrolysis of the intestinal surface, see Ugolev (1963).

Thus, for the understanding of membrane digestion it is extremely important to keep in mind that the following differences prevail in the liquid and on the phase border: (1) in the concentration of organic and inorganic substances; (2) in the pH of the medium; (3) in the energy of the molecules; (4) in the shape of the molecules; and (5) in the orientation of the molecules at the phase border.

Of no less importance are certain properties of the enzymes adsorbed on the surfaces. It has been shown, for instance, that enzymes which in a dissolved state perform hydrolysis can perform polymerization when adsorbed on the intracellular surfaces of plant cells (Kursanov, 1940; Oparin, 1959; and others).

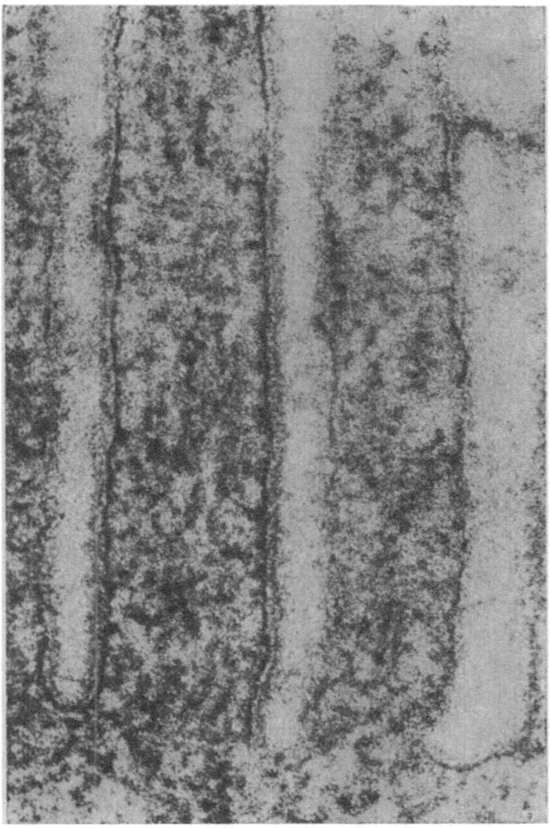

Fig. 25. Details of Fig. 24 under greater magnification. Peculiarities of the structure of the outer and inner osmophylic layers are plainly visible (according to Sjöstrand, 1963, page 531, Fig. 9).

In the review of Holden (1935) data are given on the activity of lipase which increases at the boundary of water and oil about 1000-fold compared to the activity of its aqueous solution. Deborin and Gorbacheva (1953) and Deborin et al. (1959) observed that trypsin adsorbed on organic emulsions is considerably more active than that in aqueous solution. McLaren and Babcock (1959) reported that the optimum pH of chymotrypsin adsorbed on kaolin is shifted to the alkaline side as compared to its solution in the water phase. Analogous data were obtained by comparing the activity of invertase of yeast cells and the isolated enzyme.

Characterizing the properties of the course of enzymatic reactions on the surface, McLaren attached particular importance to differences in the charge and pH on the border and inside the phases. As we have already mentioned, differences in pH may reach 1.5-2 units. There is no doubt, however, that this factor is not the only one. Apparently, of no less importance is the change in conformation of the enzymatically active enzymes as a result of fixation of the globule at definite points on the surface. Indeed, in the past year, the thoroughly investigated phenomena of allosteric inhibition and allosteric effects in general lead to the assumption that the deformation of various sections of protein molecule, including the ones farther removed from the active center, must lead to changes in specific activity. It was found (Siekewitz, 1962, cited by McLaren et al., 1963) that the enzymes bound to mitochondria are considerably more active than the same enzymes in solution. This was explained as the effects of pH (increase in local enzyme concentration), but it is not clear whether or not this settles the matter. Indeed, McLaren et al. (1963) justifiably mentions that the probability of contact of the enzyme with the substrate is greater when the enzyme and the substrate are in solution as compared with when one of the components of the reaction (in the case of membrane digestion, the enzyme) is fixed and immobile.

Spiridenkova and the author (cited by Ugolev, 1963) showed that the amylase adsorbed on the surface of the intestine is more resistant to heat and acid denaturation, in contrast to amylase in solution. Apparently, a lot depends on what surfaces and by what bonds the enzyme is adsorbed, and on the relationship of the active center to the liquid and solid phases. This problem has been insufficiently investigated, although the early investigators were aware already that the properties of enzymes adsorbed on kaolin or starch and protein are different (Holden, 1935).

Information regarding the properties of enzymes adsorbed on the surface of the small intestine is extremely scanty. The activity of amylase adsorbed on the surface of the small intestine of white rats is sharply greater than that of the dissolved enzyme (Ugolev, 1960a, 1960b). As regards the other enzymes it should be emphasized that their active centers are turned in the direction of the active phase, since adsorption of their active centers must lead to inactivation of the enzymes.

At present, when there is little doubt that many enzymes perform their functions while bound to structures of cell membranes, the following is particularly surprising. The properties of these enzymes are studied either after their preliminary solution or in systems in which enzymes are apparently present dissolved in water as well as bound to membrane fragments (the latter is the case when enzymatic activities of homogenates are investigated). This is perfectly true with respect to enzymes of the intestine proper which play such an important role in membrane digestion. In order to characterize the properties of the same enzymes in solution and on the membrane surface, Lapteva and the author conducted a series of experiments on the intestinal invertase of rats (Ugolev, 1965a). In each case the properties of dissolved invertase and the invertase of inverted pieces of small intestine of the same animal were compared. As can be seen in Fig. 26, the curve depicting the effect of pH on invertase activity of solubilized enzyme is characterized by a high peak which closely coincides with the one observed by Blair and Tuba (1963) in homogenates. On the other hand, the activity of invertase of the inverted pieces was maximal within wide limits, from 4.5 to 6.5. This fact, in our opinion, could be looked upon from two aspects, physiological and chemical. From the physiological viewpoint, the important fact is that the activity of the enzyme is retained within a wide range; this creates favorable conditions for the course of membrane digestion. From the chemical viewpoint, these data could be explained as follows. The protein globule in the adsorbed state possesses a harder structure, and its conformation is subjected to smaller changes compared to those possible when it is in solution. Comparison of our results with those of Blair and Tuba (Fig. 26) leads to the conclusion that the latter worked with preparations which contained solubilized as well as adsorbed enzymes on the membrane.

We expressed the view that conditions for the optimum activity of the intestinal enzymes are those prevailing at the surface of the small intestine rather than those found in chyme (Ugolev, 1963).

Earlier we briefly considered certain properties of the course of catalytic reactions at the phase border and the possible peculiarities of the activity of enzymes on the membrane surface in general and on the surface of intestinal cells in particular, a site at which membrane digestion takes place. Let us now look into certain matters in greater detail.

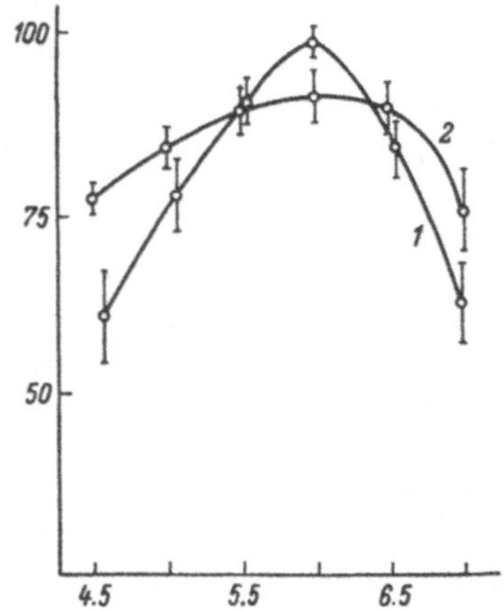

Fig. 26. Effect of pH on the activity of solubilized invertase (1) and the invertase activity of inverted pieces of the intestine (2). Abscissa: pH values; ordinate: invertase activity (in %).

Although at the present time it is well known that many important functions of the enzymes are performed when the enzymes become bound to membranal structures, often a question arises regarding the correlation of these views with no less numerous data regarding the fact that protein molecules at the phase border rapidly unfold and, as a result, lose their native structure and enzymatic functions. A series of examples of this and numerous literature references can be found in a substantial monograph by Kavanau (1965).

Apparently, a comparison of various data permits only one conclusion that, depending on certain conditions (far from always being known), the deformation of a protein molecule at the phase border could be very considerable or quite insignificant, that it may be accompanied by a decrease and loss of specific activity, or that the activity may be preserved or even increased. It is understandable that the denaturation of proteins at homogeneous interphases with high free energy is most probable, for instance at water–air or oil–water boundaries. However, even in these cases there are not a few exceptions. Indeed, adsorption of trypsin on lipid emulsions is accompanied by a considerable increase of its proteolytic activity (see the work of Deborin and his co-workers).

An even more startling example in this sense is provided by the pancreatic lipase which acquires the capacity to cleave the ester bonds of glycerides only after adsorption at the border between the aqueous and lipoid phases. According to Desnuelle et al. (1960), Braunshtein (1963), and Braunshtein et al. (1964), the appearance of activity after adsorption is a result of the deformation of the protein globule of the enzyme as a result of which the steric adequacy of the active center of the enzyme toward the corresponding substrate arises. It is of interest that pancreatic lipase in aqueous solution is not inhibited by diisopropylfluorophosphate and analogous organophosphorus compounds, but that it becomes sensitive to them after adsorption on lipids. It is important to keep this circumstance in mind, since the inhibitory effects of organophosphorus compounds are based primarily on their steric congruity toward the active center of the enzyme. Thus, if in certain cases during adsorption unfolding of a protein globule takes place which leads to destruction of its enzymatic activity, then in other cases the activity may be preserved, increased, or initiated as a result of deformation at the border of the phases.

Until now we discussed the effects of adsorption on homogeneous interphases, whereas the cell surface is of mosaic nature and adsorption on it is of a much more complex character. Indeed, if on the water–air or water–fat boundary the thread-like configuration is often energetically more advantageous, then on the membrane surface, where not only hydrophobic but also hydrophylic groupings of various charges are present, the deformation may not be connected with the unfolding of the protein globule. It is at present not clear whether in all cases the enzyme fixed on the surface of the intestinal epithelium possesses greater specific activity than the enzyme in solution.* However, for the amylase adsorbed on the surface of the small intestine, this fact, apparently, has been established by us quite clearly on THU models of the small intestine (Ugolev, 1960a) and even more convincingly by Iezuitova et al. (1964a).

In these experiments (Fig. 27) the rate of hydrolysis in vivo and in vitro during perfusion of the isolated section of the duodenum and jejunum (the access of pancreatic juice was intact) was determined. In each of these experiments several observations were made which eliminate a whole series of possible questions. The fact that the rate of hydrolysis in vivo was considerably greater than that in vitro by the enzymes which enter the intestinal cavity at the beginning of perfusion shows the presence of an auxiliary effect on the enzymes which act on the surface of the intestinal epithelium. With the desorption of the enzymes from the surface, the hydrolysis in vivo and in vitro became practically equalized. This, as we have already mentioned, serves as additional proof that the high rates of hydrolysis in the intestine are due

* This, by the way, is not a necessary condition for the course of membrane digestion. It is important that the enzyme retains certain activity in a zone where processes of hydrolysis and absorption are concomitant. From the biological point of view, the useful effect may be even greater during lower rates of hydrolysis, in cases when it is joined with absorption, than during higher rates of hydrolysis which takes place at a distance away from the resorbing surface.

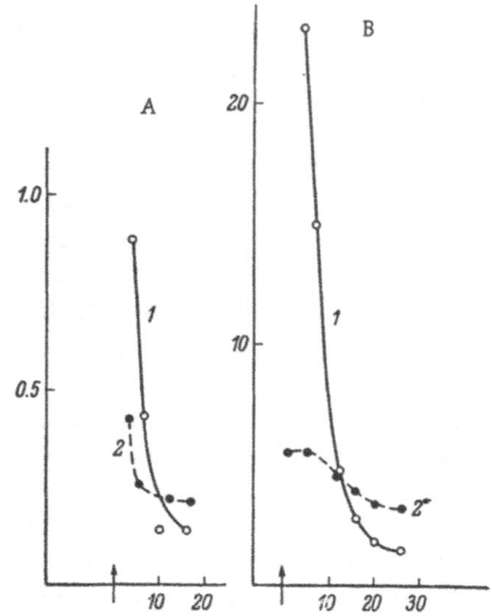

Fig. 27. Changes in the desorption of amylase during perfusion in vivo (1) and in vitro (2). (A) perfusion of jejunum–ileum; (B) perfusion of duodenum. Abscissa: time of perfusion (in minutes); ordinate: amylolytic activity (in mg of hydrolyzed starch/ml/minute). The arrow indicates beginning of perfusion.

to enzymes located on the surface of the membrane, and leaves no doubt that the calculations of the rates were sufficiently correct.

Figure 27 shows that the areas located between the curves and abscissa characterize the intensity of the enzymatic hydrolysis.

The higher rate of hydrolysis in vivo compared to total activity of the enzymes entering the intestinal cavity from the duodenum and being desorbed from the surface of the intestine can be explained only by a rise in the activity of the adsorbed amylase. It should be noted, however, that for an enzyme such as amylase with its wide spectrum of activity toward polysaccharides of various molecular weights, such a conclusion is not only too general but also not quite exact. Indeed, as we have shown earlier (Ugolev, 1960a, 1963), the increase in activity applies only to a definite spectrum of the activity of amylase, namely, its effect on the partially cleaved starch and dextrin. As regards colloidal amylase, it has been shown directly that the solubilized enzyme cleaves it faster than the adsorbed enzyme does (Ugolev, 1963). Therefore, during adsorption a change in the spectrum of activity of amylase takes place. Similar observations were made later on by several investigators. For instance, it was found that the water-soluble derivatives of proteolytic enzymes are several times less active toward proteins than toward synthetic substrates of low molecular weight (Bar-Eli and Katchalski, 1960, 1963; Cebra et al., 1961; Mitz and Summaria, 1961).

Even more striking in this respect were the data of Surinov (1965) who observed that during the binding of trypsin and chymotrypsin by derivatives of cellulose their activity may either decrease or increase (depending on the conditions of fixation of the enzymes on the carrier and its properties), but at the same time the increase in the specific amidase activity of trypsin by 30- to 50-fold takes place as compared to the initial activity of the aqueous solution of the enzyme. In all cases the maximum of activity was shifted toward the smaller molecules. Apparently, the explanation offered earlier (Ugolev, 1960a, 1963) regarding the possibility that inwardly located bonds are inaccessible to the action of enzymes fixed on a carrier is justifiable although not exhausting, because an important role must be played by factors such as deformation of the protein molecule on the surface, charge at the surface, etc. Apparently, free energy at the cell membranes is smaller than at the interphase borders of water–oil and air–water. It is therefore not surprising that the adsorption of enzymes on carriers which have polar and nonpolar groups is often accompanied by an increase in their activity. In fact, a considerable increase in the activity of chymotrypsin and ribonuclease was observed with their binding onto DNA (Diskina, 1960; Ugarova and Diskina, 1964). On complexing of trypsin and chymotrypsin (with the formation of peptide bonds) with certain forms of carboxymethylcellulose an increase in the activity of trypsin and chymotrypsin was observed (Mitz and Summaria, 1961).

Thus, membrane digestion is characterized by a series of peculiarities which depend on the fact that this process occurs at the border of various phases at which the enzyme is present on a section of a very intricately organized surface.

The immobility of enzymes fixed at the surface of intestinal cells is tantamount to a reduction in the probability of the collision between the enzyme and the substrate which is basic to the process of enzymatic reaction. Nevertheless, the effects of membrane digestion are very great. Here we are confronted with a problem whose solution is essential for the solution of many other questions in physiology and biology. In reality, according to classic views, an effective collision may be that during which several points of the surface of the enzyme, forming an active center, and several points of the substrate may be oriented in a definite manner, and bound at the active center of the enzyme. The relatively small area of these sections makes such a collision improbable. Apparently, it must be assumed that during the approach of the molecules of the enzyme and substrate their interaction is possible and as a result the final orientation of the active center of the enzyme and the reactive groups of the substrate takes place. But even if such a hypothesis is accepted, we are still too far away from the solution of the problem of how the enzymes located at the surface of intestinal cells and distributed comparatively far apart become capable of selecting the appropriate substrates.

We previously expressed the assumption that the surface of the intestinal cells, and possibly other membranes, works like "traps" (Ugolev, 1963). In other words, inert (from the catalytic standpoint) sections of the surface are capable of passing on to the active centers of the enzyme substrates which reach these sections. Since the number of inert sections of the surface may be several times greater than that of the active centers of the fixed enzymes, this could be equivalent to an equally significant increase in the effectiveness of the process, particularly at lower concentrations of the substrate in solution.

What is the possible mechanism of this process? The usual adsorption during which an exchange between the surface of the adsorbant and liquid phase (in the case of digestion) takes place cannot produce a significant effect, because hitting the inert point will not significantly affect the interaction of the enzyme with the substrate. We assumed (Ugolev, 1963) that on live surfaces under mild temperature conditions a process (mobile adsorption) may take place which occurs on the surfaces of nonbiological adsorbants at high temperatures.

Mobile adsorption or surface mobility is not a new concept in the physiochemistry of nonbiological surfaces. Briefly, the gist of the matter is that the lifetime of the molecule adsorbed on the surface may be several times longer than its retention at a given point. In other words, a molecule may slide or jump without severing itself from the surface and consequently without returning to the liquid phase and move distances which are considerably long compared with the size of the molecule. Energetically this process is determined by the fact that the energy of activation necessary for the return of the molecule into the aqueous phase is greater than the energy needed for its passage from one point of the surface to the other. Clearly, the more QA exceeds ΔQA the more intensive will be the movements of the molecule along the surface, other conditions being equal. Let us take an example provided by De Boer (1962). If QA = 10 kcal/mole and ΔQA = 5 kcal/mole, then the time of adsorption at room temperature is approximately equal to 3×10^{-6} second and the time of retention at a given point 5×10^{-10} second. Thus, during the time of adsorption the molecule makes about 6000 jumps from one place to another.

If on the basis of De Boor's work we now assume that each jump is equal to the distance between two surface atoms (about 3 Å) then during adsorption the molecule travels about 2 microns. It is clear that during this process the probability of collision of a given molecule with the active center of its enzyme increases several fold.

Since the human brain is not capable of appreciating the significance of the values discussed above, De Boor proposed an interesting model, the gist of which is the increase of the molecule of gas to the size of a bee, i.e., by about 32 million times, together with a corresponding slowdown of time. Then what occurs with gas in 1 second will occur with this model in about 1 year. Within the limits of such a model it will be evident that the bee will remain on the surface about 1.5 minutes and will fluctuate with the frequency of several thousand fluctuations per second, accomplishing the jumps every one hundredth of a second. Clearly, 6000 jumps will increase the probability that the bee will find the necessary flower.

How realistic is the two-dimensional movement on the surface of the intestinal and other membranes? Germain (1961) emphasized that at low and average temperatures the adsorption on nonbiological surfaces most frequently is localized. However, this conclusion need not frighten one since cases considered by Germain are among those in which the energy of adsorption and the energy of activation of the two-dimensional surface movement are high. Adsorption of substances on biological surfaces, as a rule, is characterized by comparatively low energies. All the evidence accumulated in biological studies indicates intensive metabolic activity by adsorbed substances between the membrane and the surrounding fluid. Apparently, the low adsorption energy of the majority of substances is a necessary condition of life. In any case, injury and destruction of cells, as a rule, are accompanied by a sharp increase in adsorption properties.

But even if we return to nonbiological surfaces, many examples can be found which show the presence of mobile adsorption at mild temperature conditions. For instance, according to the data of De Boer, the energy of activation of mobile ethane and propane on the surface of charcoal is equal to 3 kcal/mole and represents about one half the heat of adsorption. These data are of interest to compare, for instance, with the energy of activation of hydrolysis of sucrose which is equal to approximately 10 kcal/mole.

When the mobile adsorption takes place at the border of gas–solid surface the adsorbed molecules colliding with each other begin to behave like a two-dimensional gas which is in a state of equilibrium with a three-dimensional gas. The two-dimensional gas is subject to the basic laws of molecular kinetics and possesses real pressure which becomes, under equal density of condensation with the three-dimensional gas, in the first case many times greater than in the second case.

All of this is of great importance to the understanding of the processes taking place at the surface. If one imagines that at a certain point the pressure of a two-dimensional gas is lowered then a directed movement will be initiated along the gradient of pressures. Such points on biological surfaces may be the active centers of various enzymes and the pores as, for instance, in the cellular and mitochondrial membranes.

Consequently, because of the existence of two-dimensional gas and two-dimensional pressure at the surface of biological membranes, a very effective supply of the substrates can be realized toward those points of the surface where either the transformation of these substances or their flow takes place.

For the origination of two-dimensional gas it is necessary and sufficient that the energy of adsorption be higher than the energy of transposition of the molecule from a given point to a neighboring one. At present we have no direct proof of the existence of such a process on biological surfaces, but it appears very probable since the very fact of adsorption of various nutrients on the surface, particularly the carbohydrates, has been demonstrated. We mentioned earlier that the immobile classic adsorption is ineffective. Finally, there are several facts which could very well be explained by the hypothesis of two-dimensional movement of the molecules. The fact is that various surface-active substances do not affect the enzyme present in solution, have comparatively little effect on the rate of reaction at high substrate concentrations, and have a considerable inhibitory effect on the hydrolysis of sucrose by the surface of the intestine at low concentrations of this substrate.

Apparently, there are several possibilities to explain these phenomena, but the simplest one is that the enzymes themselves are not inhibited by deoxycholate and that, possibly, the so-called inert sections of the surface are blocked. At high concentrations of the substrate the rate of hydrolysis is basically determined by the direct hit by the substrate into the target and the blocking of the inert sections has no significant effect. But at small concentrations of sucrose the enzymatic activity proceeds at a high rate mainly via the passage of the substance from initial targets (from inert sections) to the final one; and now, when the initial targets become blocked, the rate of the reaction decreases to a point at which only direct hits will be the determining factors.

Figure 28 illustrates the proposed hypothesis and shows that the existence of a two-dimensional gas explains a whole series of peculiarities of kinetics on living surfaces, including (1) unexpectedly high effectiveness of the process at low concentrations of substrate, (2) a

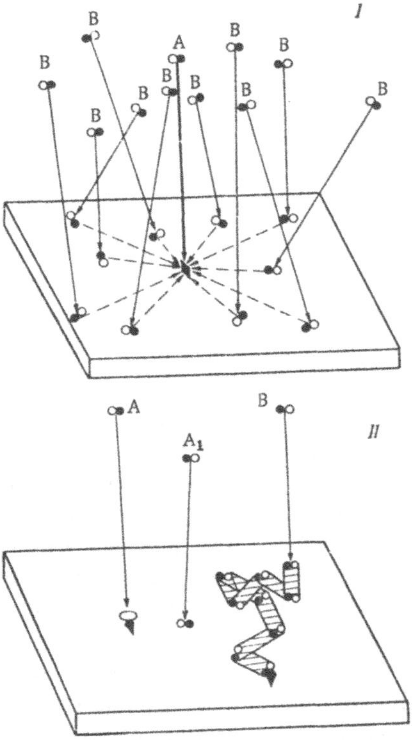

Fig. 28. Schemes illustrating the significance of two-dimensional movement along the membrane in the course of membrane digestion and other processes taking place in limited sections of biological membranes. (I) Direct (A) and secondary (B) hits of substrate on the target possessing enzymatic or transport functions (dark diamond); (II) direct hits on the active center (A) and inactive section of the membrane (A_1), (B) hit on the active center of the molecule of the substrate which moves or jumps along the surface.

higher temperature coefficient compared to that with dissolved enzyme, (3) nonspecific inhibition during blocking of inert sections, and so forth.

We will now attempt to review anew the biological aspects of this hypothesis which includes the assumption of the existence of primary targets and the presence of two-dimensional movement of substances on the surface. The high rates of metabolic processes may be accomplished by other means, for instance, by the increase in the concentration of enzymes on the surfaces or by an increase in the concentration of substrates which enter from the solution. However, there could not be many enzymes of one kind on the surface for the simple reason that on the surface tens of various kinds of enzymes, pores connecting both sides of the membrane, etc., must be distributed. It is characteristic of biological systems that at each given moment the concentration of substrates is comparatively small.

It is possible that nature, instead of increasing the area of the targets at the expense of the increase in the concentration of enzymes, accomplished the same effect by the creation of primary receptors or primary targets which do not directly participate in the process of transformation of substances, are not highly specialized or specific

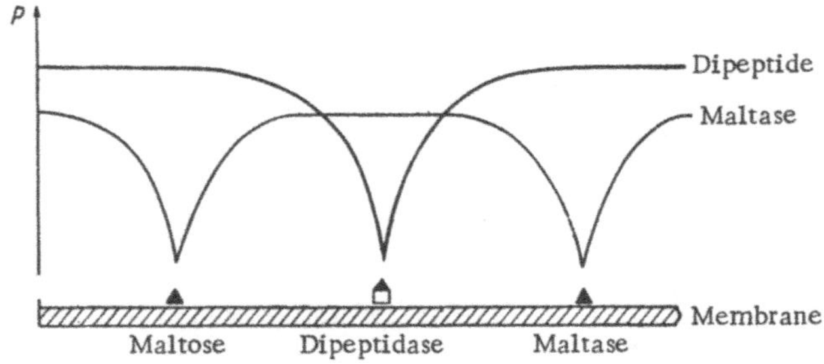

Fig. 29. Possible distribution of partial pressures in two-dimensional gas on the membrane surface, definite points of which possess dipeptidase or maltase activity. P is the pressure of binomial gas.

for one or a limited concentration of substrate, but ensure the high effectiveness of the performance of a few enzymes on the surface by furnishing them with the mobilely adsorbed substrates. As can be seen from Fig. 29, the direction of the movement of the substrates toward their targets may be a result of the gradient of the partial pressures in the two-dimensional gas. It is because of this circumstance that flow inside the cell is unerringly found for the carbohydrates and neutral, basic, and acidic amino acids. For purely physicochemical reasons sucrose will be moving toward a point possessing invertase and not peptidase activity. If the concentration in the two-dimensional gas is not too great then each type of molecule will be moving in strict accord with gas laws practically independently of each other.

This hypothesis, as we can see, explains not only certain regularities of membrane digestion but also the processes of active transport. In addition, we believe, it could be useful to explain how the substances find their receptor targets. Finally, it permits the assumption that there exist in the cells not only canals but also a peculiar highway along which the molecules glide or jump in definite directions. One must not forget, however, that at the base of this hypothesis are certain physical regularities which are possibly active within the cell and certain facts which could have a different explanation and which should not be considered as direct proof of the described hypothesis.

This, however, has been done not because attempts to find convincing proof have not been made.

The characterization of membrane digestion as heterogeneous catalysis on porous dissimilar surfaces would not be complete if we did not take into consideration the fact that we are considering a living porous catalyst whose properties may be regulated by many mechanisms in several respects.

At the present time we know very little about this. It has been shown that large concentrations of bile inhibit the adsorption of amylase on the intestinal surface (Lee, 1962), while small concentrations of bile stimulate it (Ugolev, 1960b, 1963). Palay and Karlin (1959a, 1959b), on the basis of their electron microscopic investigations, proposed that microvilli are capable of contraction and relaxation and this certainly will be reflected in the course of membrane digestion.

We have already pointed out (Ugolev, 1963) that the size of the pores of the brush border markedly affects the course of catalytic processes on the surface of the microvilli. At that time a hypothesis was expressed that the sizes of the microvilli undergo changes due to their contraction or relaxation and, therefore, in contrast to nonliving porous catalysts, this important parameter which affects membrane digestion is changeable in living organisms. Shortly afterward it had been shown that contractile proteins are essential components of various cell membranes (see reviews by Green, 1964; Neifakh, 1964; Poglazov, 1965). Of particular interest are the data on the presence of myosin-like protein in various cell membranes, including such typical noncontractile cells as the cells of rat and mouse liver (Neifakh, 1964; Neifakh and Vasilets, 1964).

Contractile structures of membranes may have direct bearing on the regulation of permeability of cell walls and also on the regulation of space relationships between the various components of enzyme ensembles (for instance, bringing the enzymes closer together) (see reviews by Green, 1964; Neifakh, 1964). Until now this was clearly related to mitochondrial functions. During contraction of the contractile proteins on the surface of microvilli membranes, one could expect a decrease in their diameter and, consequently, an increase in pore size. In addition, Millington and Finean (1962) and later many other investigators confirmed the data of Goldin (1956) and Shestopalova et al. (1960) on the presence of core in microvilli. By means of large magnifications it has been shown that the core is not a canal but a bunch of fibrils. It is assumed that fibrillar structures may perform the supporting and contractile functions.

Thus we know at present a little more than we knew a few years ago regarding the manner in which the contraction and relaxation of microvilli takes place. However, a decisive analysis is still ahead. Nevertheless, as has been previously mentioned (Palay and Karlin, 1959a), even in 1959 it had been shown that the sizes of microvilli may undergo considerable changes in short intervals of time after feeding, for instance, which suggests that this phenomenon is real.

Chapter IV

Detailed Characterization of Membrane Digestion in Higher Animals

In this chapter we dwell in greater detail on such problems as the interrelationship of membrane digestion with cavital digestion and absorption, on the possible role of the motor function of the small intestine in the realization of enzymatic hydrolysis on the membrane surface, and others. At the same time certain problems will not be discussed to avoid repetition. It must be kept in mind that we are describing the membrane digestion of mammals, but it must be understood that many regularities are sufficiently general and, therefore, of significance toward the understanding of the hydrolytic processes on cell membranes of other classes and even species of animals.

INTERRELATIONSHIPS OF CAVITAL AND MEMBRANE DIGESTION

Let us evaluate the role which cavital and membrane digestion plays in the gastrointestinal tract of higher animals and man. Two hypotheses are possible: (1) cavital digestion and membrane digestion are duplicate mechanisms, that is, mechanisms which perform a similar or identical function, or (2) functions which are performed by both types of digestion are not identical and each mechanism in this sense is irreplaceable.

If the first hypothesis is correct then it could be expected that the assemblage of enzymes acting in the intestinal cavity and on its surface should be about the same. However, cavital digestion is performed by the enzymes produced by the gastric and pancreatic glands (pepsin, gelatinase, parapepsins, possibly cathepsins of the stomach, trypsin, chymotrypsin, carboxypeptidase A, carboxypeptidase B, amylase, ribonuclease, deoxyribonuclease, and lipase which mainly cleaves tri- and diglycerides but not monoglycerides). Aminopeptidase, leucine-aminopeptidase, tri- and dipeptidases, prolinase, nucleotidases, nucleosidases, trehalase, invertase, maltase, lactase, monoglyceride-lipase, and others act on the intestinal surface.

It is not difficult to see that primarily the initial stages of hydrolysis of proteins, carbohydrates, fats, and nucleic acids are performed in the cavity and the enzymes of the small intestine are responsible for the final stages of hydrolysis. Even on the basis of comparison of the enzyme composition of the intestinal content and on the intestinal surface one can make a far-reaching conclusion regarding the interrelationship of cavital and membrane digestion. The initial stages of the hydrolysis of nutrients is performed by means of cavital digestion, while the intermediate and final stages of hydrolysis take place on the intestinal surface.

A series of other facts support this. Membrane digestion takes place on the surface of microvilli, the distance between which varies from 100 to 200 Å. Apparently, because of this circumstance, the molecules whose size exceeds the diameter of the pores of the brush border cannot penetrate the brush border and the membrane digestion will not be effective there. Since the majority of animals utilize the tissues of multicellular and unicellular organisms as food, for these animals cavital digestion is an absolutely essential initial stage in the utilization of food.

In the subsequent section where the topography of membrane digestion is considered, it is demonstrated that the stomach mucosa of omnivorous animals does not give indications of the existence of this mechanism there. Apparently, the stomach in the course of evolution arose as an organ which stores food and in which the initial stages of cavital digestion are performed.

Food entering the small intestine from the stomach even after preliminary cleavage of about 10% of the peptide bonds and the denaturation of the cell structures is still unavailable to the action of enzymes which are bound to the structures of the intestinal cells.

At this stage of digestion the effectiveness of the enzymes acting distally is particularly evident. As has been shown by the early investigations of Northrop (see reviews by Northrop et al., 1950), proteolytic and other enzymes easily penetrate the injured walls of tissues and therefore are capable of hydrolyzing substrates which are located inside the cells. The comparatively small molecules thus formed penetrate beyond the cell limits and reach the surface of the mucosa and are subjected to hydrolysis in the zone of the brush border.

Thus, cavital digestion is the most effective mechanism for the hydrolysis of large nutrient molecules, supermolecular aggregations, and particularly of cellular material, while membrane digestion is effective mainly toward the intermediary products of hydrolysis.

Direct experiments conducted by us (Ugolev, 1960b) showed that colloidal polysaccharides (amylose and glycogen) are practically not hydrolyzed at the surface of the intestinal wall in contrast to the products of their partial hydrolysis. The latter circumstance appears particularly important and needs further consideration. The hydrolysis of polysaccharides in the cavity as well as on the surface of the intestine is performed by pancreatic amylase. For this reason one could expect that the effects of the same enzyme should be similar. However, membrane digestion is performed by enzymes fixed at the surface of the membrane and, therefore, only those bonds will be hydrolyzed which, because of the geometric peculiarity of the substrate molecule, are accessible to the action of the active group of the enzyme. The adsorbed enzyme cannot be oriented toward the substrate, neither can it penetrate the substrate. Earlier we mentioned that the enzymes acting in the cavity are, as a rule, endohydrolases. The intestine proper enzymes are exohydrolases. We should now make the important conclusion, namely, that the same enzyme molecule possesses endohydrolytic properties inside the phase and exohydrolytic ones in the adsorbed state.

Thus, cavital digestion is a mechanism for the initial stages of hydrolysis, while membrane digestion is a mechanism for the intermediary and final stages of hydrolysis. If so, then where is the border which demarcates the passage from cavital to membrane digestion? At the present time one can answer this question only approximately. The hydrolysis in the brush border is possible only when the sizes of the substrates' molecules become smaller than the sizes of the pores of the brush border (at least in relation to water-soluble substrates). As an example, the molecule of myoglobin (molecular weight 17,000) may have the shape of an irregular globule of the sizes 40 × 35 × 25 Å. Thus, the majority of pretty large molecules can penetrate the zone of the brush border.

However, purely geometrical considerations are insufficient to solve this problem. Indeed, we have already mentioned that an important factor limiting the effectiveness of membrane digestion is the availability of the hydrolyzable bonds to the enzyme fixed at the surface.

It is at present not clear to what extent membrane digestion is effective toward proteins which retained their secondary and tertiary structure. On the other hand, it is important to keep in mind that at the border of phases rapid denaturation of proteins as well as changes in the configuration of the polymers take place. Certain regularities in the transition from cavital to membrane digestion were investigated by Ugolev and Marauska (1964) employing hydrolysis of soluble starch.

Acute experiments were performed on white rats weighing 150-160 grams under nembutal anesthesia. The isolated upper and caudal sections of the small intestine of 20-25 cm length were perfused, employing the solution of starch–Ringer (the concentration of soluble starch was 0.075%, pH 7.0) at the rate of 2-5 ml per minute. The collected perfusate was rapidly cooled to interrupt the enzymatic reaction and then diluted (1 ml of perfusate plus 5 ml of water plus 0.5 ml of 1 N HCl). To obtain the iodine–starch complexes, 0.5 ml of a solution of 0.1% iodine and 1% KI was added to the test solutions thus prepared. The controls were similarly prepared, but instead of the perfusate 1 ml of a 0.075% solution of starch was used (taken as 100%). Optical density was determined spectrophotometrically (spectrophotometer type SF-4) in the range of 400 to 600 millimicrons at intervals of 20 millimicrons. In some experiments the optical density of the iodine–polysaccharide complexes was determined by a recording spectrophotometer type SF-2M.

In order to investigate the hydrolysis in vitro, the collected perfusate was incubated in a water ultrathermostat at 38°C until the iodine reaction was negative. An equal volume of an 0.15% solution of starch (the concentration of the latter in solution in vivo and in vitro was thus equal) was then added to the perfusate and incubated in the ultrathermostat. One ml of solution was taken from the incubation mixture every 1-2 minutes and added to a test tube which contained 5 ml of water, 0.5 ml of 1 N HCl, and 0.5 ml of I_2 in KI. Spectrophotometric measurements were then performed.

Since the changes in optical density during the hydrolysis of starch are a result of a decrease in the initial substrate as well as a decrease in the formation of various dextrins, it became essential to differentiate these products. This could be accomplished by comparing the curves

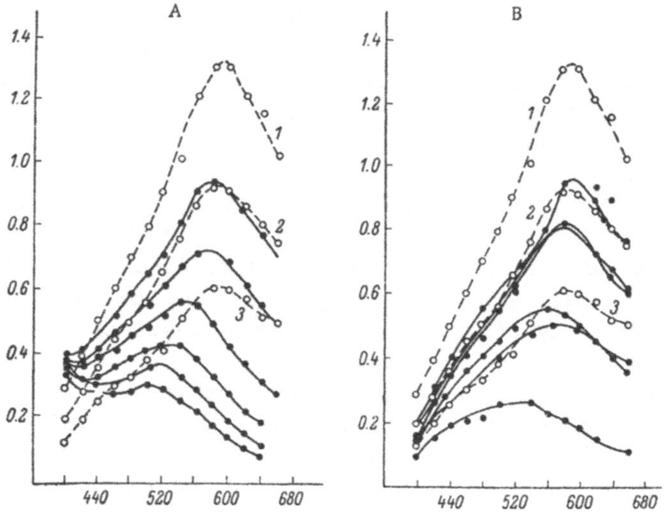

Fig. 30. Absorption spectra of iodine–starch complexes at various stages of cleavage of soluble starch in vitro (A) and in the intestine (B). Abscissa: wavelength (in millimicrons); ordinate: optical density (in apparatus SF-4). Dotted lines: absorption spectra of intact soluble starch. (1) 0.075%; (2) 0.0562%; (3) 0.0375%.

of utilization of starch in the course of its hydrolysis with the curves for intact starch at various concentrations.

As can be seen from Fig. 30, the absorption spectrum of substrate–starch complexes in our experiments was similar to that described by other investigators (Swanson, 1948; Kerr, 1956). The maximum of optical density corresponded to the wavelength of 580-600 millimicrons; comparison of the solutions of various starch concentrations shows that a linear relationship exists between the concentration of the substrate and the optical density.

During the enzymatic cleavage of starch in vitro by enzymes which enter the perfusate the results obtained were similar to those obtained earlier. In the course of cleavage (Fig. 30A) a reduction in the extinction at 580-600 millimicrons and the displacement of the maximum toward shorter wavelengths takes place. Particular attention should be directed toward the relative increase in the optical density at wavelengths shorter than 580 millimicrons as compared with that corresponding to the content of starch in a given test solution. As we mentioned earlier, these phenomena reflect the accumulation of dextrins in the course of hydrolysis of starch.

Significantly different results were obtained in the determination of absorption curves for the starch hydrolyzed in vivo. As can be seen from Fig. 30B, the curves in this case were similar to those for the controls. A significant shift of the maximum was obtained only after hydrolysis of more than 60-70% of the initial substrate. As a comparison of the optical densities of the experimental and control test solutions shows, during hydrolysis in the intestine the accumulation of dextrins occurs in insignificant amounts. The relationship between the extent of hydrolysis of starch and the accumulation of dextrins can be readily examined by comparing the optical density of the control (intact starch) and experimental (starch cleaved in the test tube or in the intestine) test solutions at wavelengths which are characteristic of iodine–starch and iodine–dextrin complexes (Fig. 31). The increase in optical density at 460 millimicron wavelength, as compared with the line for control, reflects the accumulation of the red component (mixture of dextrins), which permits the following conclusions.

During hydrolysis in a test tube a remarkable accumulation of dextrins takes place, which increases in definite limits as the hydrolysis proceeds. During hydrolysis in vivo the accumulation of dextrins is small. This effect depends partially on the fact that hydrolysis in vivo proceeds faster than in vitro. As can be seen from Fig. 32, during the cleavage of soluble starch in a test tube the accumulation of dextrins is greater the lower the content of amylase. At first sight all the differences between hydrolysis in vivo and in vitro could be

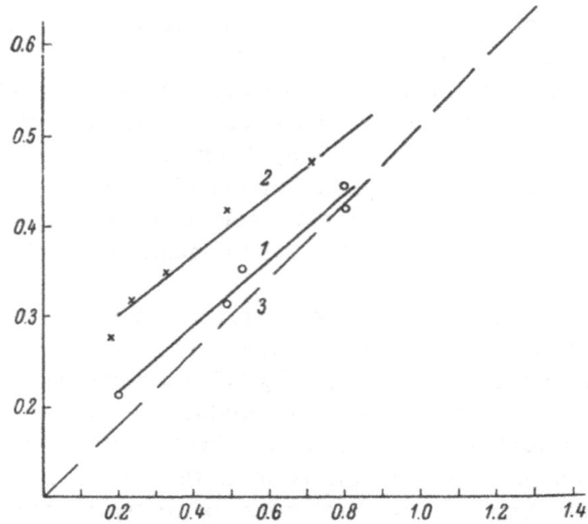

Fig. 31. Relationship of optical densities of iodine–polysaccharide complexes at different stages of hydrolysis of starch in vivo (1) and in vitro (2), and of uncleaved soluble starch (3, control). Abscissa: optical density (in numbers of the scale of SF-4 at wavelength of 580 millimicrons mainly reflecting the content of starch); ordinate: same, at the wavelength of 460 millimicrons, indicating mainly the content of dextrins.

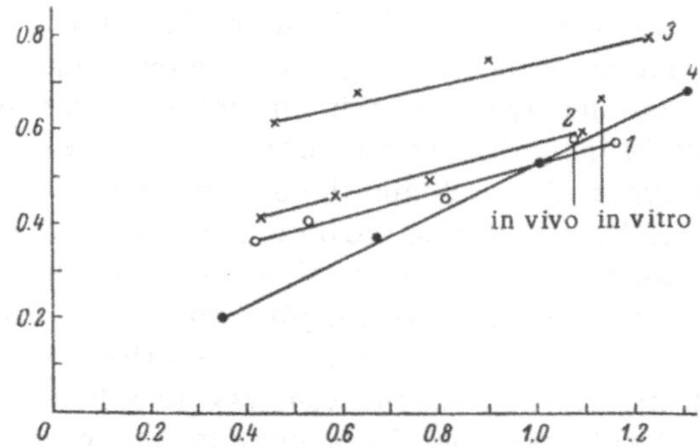

Fig. 32. Accumulation of dextrins at different amylase concentrations [(1) undiluted perfusate; (2) threefold diluted perfusate; (3) tenfold diluted perfusate; (4) control] and during equalized rates of hydrolysis in vivo and in vitro. The rest of the notations are the same as in Fig. 31.

reduced to the difference in the rate of hydrolysis. With closer examination, however, this conclusion would appear inadequate. As is known, during prolonged perfusion of rat intestine the desorption of amylase from the surface of the intestine and a slowdown of hydrolysis to almost complete cessation take place. Were the peculiarities of cleavage of starch exclusively dependent on the rate of the enzymatic reaction, the absence of dextrins should have been observed only during the initial periods of perfusion. In reality, however, even at the end of the experiments, when the rate of hydrolysis in vivo is several times smaller, one could not detect in the perfusate the accumulation of dextrins. Moreover, in special experiments in which the hydrolysis of starch in the intestine was compared to that in the test tube at equal reaction rates and equal extents of hydrolysis, definite differences were retained (Fig. 32).

Since the intestinal wall is impermeable to dextrins, or only slightly permeable, a conclusion suggests itself that the intermediary products of hydrolysis are eliminated to the brush border where they are cleaved and absorption begins. Analysis of samples of starch cleaved in the intestine or the test tube leads to the conclusion that the greatest difference is in the region of the wavelength of 480-460 millimicrons and shorter. This leads to the assumption that membrane digestion is particularly related to the cleavage of dextrins, beginning, for example, with 10 glucoside residues or less, although we have previously shown the possibility of cleavage of soluble starch at the surface of the intestine.

Confirmation of these data employing other methods leads to the conclusion that more than 80% of the glucoside bonds are hydrolyzed in the zone of membrane digestion. Also, the data obtained help us to understand one of the mechanisms which increases the effect of membrane digestion on the initial stages of hydrolysis which take place in the small intestine. During hydrolysis in the test tube the amylase is distributed between starch and the intermediary products of its hydrolysis; but since during digestion in the intestine the latter do not accumulate in the chyme but enter the brush border, a supplementary effect arises as a result of this, since now the amylase acts primarily on the initial substrate.

The approach described in this work may prove useful for the diagnosis of the state of cavital and membrane digestion. Let us by this example attempt to give a better description of the relationship between cavital and membrane digestion.

When soluble starch (polyglucoside chains consisting of 30-35 glucoside residues) serves as the nutrient substrate from the very beginning the hydrolysis takes place in the cavity and on the surface. After the formation of dextrins consisting of 8-12 residues, that is, after cleavage of two or three glucoside bonds, the process is completed entirely in the brush border.

Thus, in the cavity at the most 3 and in the zone of membrane digestion 26-30 bonds (approximately in the ratio of 1 to 10) are hydrolyzed. The role of cavital digestion thus appears insignificant and the mechanism itself appears to be auxiliary. One must keep in mind, however, that starch is supplied in the form of granules which are not available to membrane digestion. Thus, before the dextrins consisting of 20-30 glucoside residues are formed, the destruction of the starch granules as well as the hydrolysis of colloidal starch is necessary [for instance, the molecule of amylose contains about 3800 glucoside residues (Honeyman, 1962) and, therefore, about 100-200 glucoside bonds must be disrupted].

It can be seen that the role of cavital digestion is quite essential. However, during the assessment of the relative significance of both hydrolytic processes it becomes clear that 70-80% of the glucoside bonds or, perhaps, even more are cleaved in the zone of membrane digestion.

Although at present we do not know the zone of transition from cavital to membrane hydrolysis, it is clear that sufficiently large peptides can penetrate the brush border, and that the hydrolysis of the peptide bonds is also completed in the zone of the brush border.

DESQUAMATION OF THE INTESTINAL EPITHELIUM AND ENZYMATIC HYDROLYSIS IN THE CAVITY OF THE SMALL INTESTINE

We should consider still another viewpoint which, apparently, has only historical interest but is, nevertheless, discussed in modern literature (Shlygin, 1964). According to this viewpoint, the final stages of hydrolysis occur in the gastrointestinal tract under the influence of enzymes of exfoliated epithelium.

About 100 years ago, Leube (1868) drew attention to the fact that the solid portion of the intestinal juice contains the bulk of the enzymes of the cavity. He wrote: "These pieces are normal component parts of the intestinal juice. They were always present in material separated from fistulas and consist of cells bound by a mucin-free intermediate substance. Thus, it could be stated that the places from which the enzyme originates are these flocculent bodies; from here the enzyme is distributed throughout the fluid" (page 291).

At the present time this point of view is developed further by Fomina (1951, 1957) and Shlygin (1952, 1958). The essential points of Leube's assumption remain practically unchanged. Thus, in his last review Shlygin (1964) wrote: "Glandular cell walls of the intestine form and accumulate enzymes and at a definite stage of their activity they are discarded into the intestinal lumen where, upon disintegration, they furnish the enzymes to the surrounding fluid" (pages 21-22).

Before we cite certain important data of investigators who do not attach much importance to the mechanism of defoliation in the course of the processes of digestion of food, we shall first examine the essential arguments of the defenders of this point of view.

Lately, however, Shlygin departed from his original opinion and he now considers that only enterokinase acts in the cavity of the gastrointestinal tract. As regards the other enzymes (phosphatases, peptidases, disaccharidases), the viewpoints have been substantially altered. According to Shlygin (1964), peptidases and the intestinal lipases act only intracellularly. Alkaline phosphatase and disaccharidases remain. These enzymes, in accord with the new compromise views of the adherents to the theory of defoliation, possess double activity, that is, they participate in hydrolysis when the cells are still the component parts of the villi and when they are defoliated and destroyed in the gastrointestinal lumen. It remains only to understand what the interrelationship of the effects of alkaline phosphatase and disaccharidases on the intestinal surface and in chyme is.

Even if it is assumed that during the destruction of cells the activity of these enzymes remains unchanged, there is no ground whatever to consider cavital and membrane effects as identical.

Indeed, during the comparison of the activity of alkaline phosphatase and invertase in equal volumes of the scrapings of the mucosa and exfoliated cells, only the activity per unit cellular material is being determined. The total extent of hydrolysis can be calculated only by taking into account the total amount of enzymes which entered the cavity and which act on the surface. Such a recalculation leads to opposite results. Indeed, if the rate of renewal of the intestinal epithelium of the small intestine is about 3 days, then about 1.5% of the intestinal epithelium is desquamated per hour (for details see Table 2). In other words, the distribution of enzymatic potentialities between the surface of the intestine and its content with respect to disaccharidases and phosphatases is about 98.5:1.5. The power of the mechanisms, therefore, differs by about 2 orders of magnitude.

The discarded epithelium is quite rapidly moved to distal sections of the small intestine and, therefore, had this mechanism been of any material significance, one could expect that the final stages of digestion should be accomplished in the distal sections of the small intestine and be least in evidence in the proximal regions. As is well known, in reality the resorption of the nutrients is concluded predominantly in the uppermost sections of the small intestine (see reviews by Wilson, 1962; Ugolev, 1963; Isselbacher and Senior, 1964).

TABLE 2

Time of Replacement of Intestinal Epithelium

Section of the intestine	Object of studies	Time of replacement		Rate of exfoliation of epithelium per hour (in %)*	Literature sources
		Days	Hours		
Duodenum	Rats	1.57	–	2.6	Leblond and Stevens, 1948
	Rats	–	63	1.6	Warburton, 1961
	Mice	<2	–	2.08	Leblond and Walker, 1956; Leblond and Messier, 1958
	Cats	2.3	–	1.8	McMinn, 1954
	Man	1.8	–	2.3	Bertalanffy and Nagy, 1958
Jejunum	Rats	1.4	–	2.6	Widner et al., 1951
	Rats	–	52	1.9	Warburton, 1961
	Mice	1.8	–	2.3	Knowlton and Widner, 1950
	Mice	3	–	1.4	Leblond and Walker, 1956
Ileum	Rats	1.4	–	2.6	Leblond and Stevens, 1948
	Cats	2.8	–	1.5	McMinn, 1954
Small intestine	Mice and rats	2-3	–	2.08-1.4	Hooper, 1956, 1962
	Man	3-6	–	1.4-0.7	Lipkin, 1965

*Time of complete replacement of epithelium is taken as 100 (recalculation is ours).

The significance of the activity of the enzymes acting in the cavity is indicated by the recent data of Dahlqvist and Borgström (1961). They found that the activity of lactase in the cavity of the small intestine is so low that it required about 100 hours (more than 4 days) for the complete hydrolysis of the introduced lactose. In fact it is hydrolyzed and completely absorbed in 3-4 hours. Approximately similar data are available regarding the distribution of the enzymatic effects of intestinal hydrolases between the cavity and the surface.

As is shown in the next chapter devoted to the pathology of the gastrointestinal tract, if the rates of the final stages of hydrolysis were materially dependent on the enzymes of the exfoliated epithelium, then we would be confronted with the most severe forms of food intolerances.

The theory reviewed originated 100 years ago when it appeared impossible to imagine that intestinal enzymes can act as components of living cells. At the present time it is clear that they can do so and, moreover, ample data are available to show that they are doing just that.

The completion of hydrolysis by living cells has other essential advantages. This process is accomplished under sterile conditions (see pages 95-98) and it is in close contact with the processes of resorption.

It appears more probable that in the course of exfoliation the transfer of enterokinase into the intestinal cavity which activates trypsinogen takes place. We should like to express the hypothesis, however, that this process may be effectively accomplished by the contact of trypsinogen with the surface of the small intestine.

Thus, at the present time there are no data in support of the contention that the enzymes of desquamated intestinal epithelium play any significant role in the course of digestive processes.

THE EFFECT OF MEMBRANE DIGESTION ON THE INITIAL STATES OF THE HYDROLYSIS OF POLYMERS

In the earliest experiments in which the effects of pieces of small intestine of rats on the rate of hydrolysis of starch had been studied, the fact had been observed which could be interpreted only as an increase in hydrolytic processes in the incubation medium by the enzymes which act on the surface. If we denote the rate of hydrolysis of starch by a certain amount of amylase solution as V_1 and the rate of hydrolysis of starch during incubation with an intestinal slice as V_2 then during simultaneous incubation one could expect that the rate would be $V_1 + V_2$ in the presence of excess substrate but less than $V_1 + V_2$ if the concentration of the substrate is not sufficiently high. The interaction of cavital and membrane digestion, it would appear, is not simply a summation of the enzymatic processes. We have already discussed this in the preceding section, but at that time we discussed the increase in the rate of appearance of the end products of the reaction. Here, however, we should like to discuss experiments in which initial stages of hydrolysis were determined (Ugolev, 1960a, 1960b, 1960c; Ugolev et al., 1961; Iezuitova et al., 1964a).

How can the stimulation of the early stages of hydrolysis occur if the initial substrates do not penetrate the brush border and the enzymes which act on the surface are not in solution?

Let us consider an ideal case when a molecule of amylose consisting of 100 glucoside residues is hydrolyzed while after each cleavage of one glucoside bond two equal and almost identical molecules are formed. The first product will contain 50 glucoside residues, the second 25, the third 12 and 13, and the fourth 6 and 7.

The molecules of amylase are approximately related to the initial substrate and to all the products of its hydrolysis up to pento- and hexoglucosides (Honeyman, 1962).

The distribution of the enzyme between the initial product and the products of the reaction at any given moment will be determined by the number of targets (their number can be ignored if the concentration of substrate in solution is not too great).

As an example let us examine the distribution of the enzyme in solution consisting of a certain number of amylose molecules (consisting of 100 glucoside residues), an equivalent amount of the first product of the reaction (polyglucoside of 50 glucoside residues), the second product (polyglucoside of 25 glucoside residues), and the third product (of 12 and 13 glucoside residues). If the enzyme is distributed proportionally to the number of targets (earlier it was noted that the relationship of the enzyme to all these substrates is equal), then the relationship will be 1:2:4:8. From this the following conclusions can be drawn: (1) about 7% of the amylase molecules in the solution will act on the initial substrate of the mixture; and (2) the third product will bind more than half the entire amount of the enzyme.

If one now recollects that the dextrins which contain more than 10 glucoside bonds are not detected in the intestinal cavity and, as has been shown earlier, are completely cleaved in the zone of the brush border, then it becomes evident that the rate of the initial stages of hydrolysis must sharply increase.

In reality the effect of membrane digestion on the initial stages of hydrolysis taking place in the cavity of the gastrointestinal tract will be even greater. Firstly, in the zone of the brush border, molecules consisting of 30 or more glucoside residues are cleaved. Therefore the number of targets which are attacked by amylase acting in the small intestinal cavity is sharply decreased at the expense of the intermediary products of hydrolysis. Secondly, we are not taking into account the possibility of the inhibitory action of oligosaccharides, although it is well known (Honeyman, 1962).

Thus, because the products of partial hydrolysis are continuously eliminated from the zone of the brush border one observes an intensification of the initial stages of cleavage of starch, although it is clear that membrane digestion does not directly participate in this process.

In the language of cybernetics, a positive mutual contact is established between the cavital and membrane digestion which ensures maximal effectiveness of both mechanisms: the cavital digestion prepares the initial substrates for membrane digestion; and membrane digestion reduces the cycle of the treated substrates by the passage of the products of partial hydrolysis into the brush border. The analogous mechanism, probably, is of significance for the stimulation of the initial stages of hydrolysis of proteins, although the zone of passage and, hence, of effectiveness of the influences of membrane digestion on the course of cavital digestion in this case have not as yet been investigated.

TOPOGRAPHY OF MEMBRANE DIGESTION

We have already mentioned several times that membrane digestion in higher animals occupies an intermediate place between cavital digestion and absorption. This mechanism ensures the intermediate and final stages of hydrolysis as well as the transition from digestion proper to absorption. For this reason it is understandable that in higher omnivorous animals membrane digestion occurs mainly in the small intestine.

In special experiments we (Ugolev, 1960a, 1960b, 1960c) employed trichloroacetic acid (TCA)-treated sections of stomach, small and large intestine, and esophagus mucosa of white rats and showed that the "membranal" phenomena of amylase are sharply revealed only in the small intestine and that they were practically absent in the esophagus, stomach, and large intestine.

We also pointed out that the enzymes performing the hydrolysis of oligomers are produced mainly in the intestinal epithelium and are localized on its surface directed toward the cavity of the small intestine. This is an important argument in favor of the postulate that membrane digestion in higher omnivorous animals is performed in the small intestine.

Let us now examine in greater detail the topography of digestive processes in the small intestine.

Several authors (Fisher and Parsons, 1953; Borgström et al., 1957; McGeachin and Ford, 1959; Dahlqvist and Borgström, 1961; Spencer, 1964a; and others) observed that the concentration of enzymes in the small intestinal cavity decreases from the duodenum in the caudal direction. Thus, sufficient data are available at present to consider that cavital digestion in the small intestine has a proximal–distal gradient.

The topography of membrane digestion is somewhat different. It can be assessed on the basis of the distribution of the enzymatic activity in various sections of the small intestine.

During the perfusion of various sections of the small intestine with soluble starch it was observed that membrane digestion was sharply revealed in the duodenum (De Laey and Iezuitova, 1962; Iezuitova et al., 1964a). Maximal effects were observed in the upper sections of the jejunum and practically absent in distal sections of the ileum (Fig. 33).

Table 3 contains data which characterize the proximal–distal gradient of the digestive and resorptive functions in various animals. It should be mentioned that these data are not free of contradictions, since the distribution of enzymes along the intestine changes depending on the state of nutrition, age, and species of animals. In addition, the differences may depend on the methods employed in the enzymatic studies (use of homogenates and intact sections, large or small slices of the intestine).

Certainly of the greatest interest is the distribution of enzymatic activity between the contents of the small intestine and its surface. Such data were obtained by Iezuitova, Timofeeva, Nurks, and the author on adult fed rats. The enzymatic activity was determined one hour after feeding. As can be seen from Fig. 34 (small intestine was used without duodenum), the greatest amylolytic activity in regard to membrane hydrolysis of starch is observed at the beginning of the jejunum and the least is in the end of the ileum. On the other hand, the concentration of amylase in the intestinal contents markedly rises in the

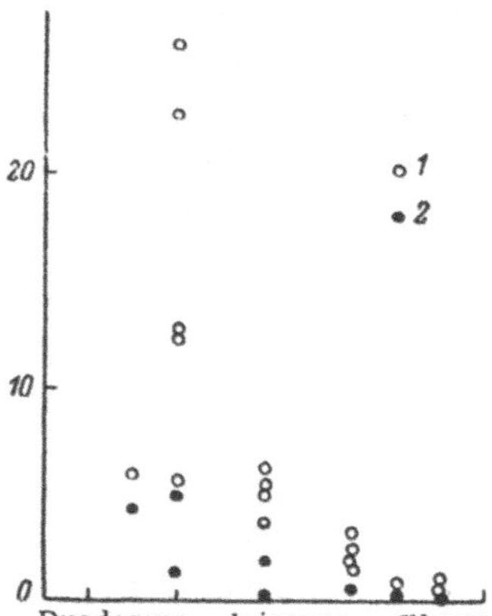

Fig. 33. Membrane and cavital hydrolysis of soluble starch in vivo (1) and in vitro (2) in various sections of the intestine of white rats. Abscissa: sections of the intestine; ordinate: amylolytic activity (in mg of hydrolyzed starch/ml/minute).

TABLE 3

Spatial Distribution of Various Intestinal Activities (According to Spencer, 1964a, pp. 286-288)

Activity	Object of study	Notes
Adenosinetriphosphatase	Hamster	Insignificant increase in the first section of ileum
	Rat	Approximately equal distribution
Alkaline phosphatase	Rat	Exponential decrease
	Mouse	Decrease approaches exponential
Amylase	Rat	Concentration in jejunum is slightly higher than in duodenum
Transport of aromatic amino acids	Hamster	Distribution approaches normal: $\alpha = b \exp[(L_0 - L)^2 C]$ where L_0 is the distance from middle point of the intestine to pylorus and C is the final concentration
Absorption of bile salts	Rat	Close to $R = mL^2 + b$
Absorption of calcium	Rabbit	Approximately $(\alpha - B)(L - D) = c$; B, D, c are constants
	Rat	Absorption is active in 10-15 cm of proximal section only in vitro, but could change in vivo
Chitinase	Lombricus	More in middle portion of the intestine than in distal
Cholinesterase	Rabbit	Higher in ileum than in duodenum
Dipeptidase	Rat	Higher in ileum than in other sections
	Locust	Anatomical localization in the middle section of the intestine was not noted
Disaccharidases	Rat	Invertase higher in 3-5 sections from 7
	Pig	Invertase, maltase, isomaltase are localized mainly in the distal third of the intestine, whereas trehalase, lactase, and cellobiase are in the proximal third
	Man	Lactase is absorbed more proximally than maltose, and the latter more proximally than sucrose

TABLE 3 (continued)

Activity	Object of study	Notes
Movement of electrolytes	Dog	Sodium enters duodenal contents and gradually appears in the ileum contents during adsorption of mannitol
"Erepsin"	Cat	Most in duodenum; concentration in jejunum is $^2/_3$ of that in duodenum and $^1/_2$ of that in ileum
Absorption of fat	Rat	In jejunum, but in large doses also in ileum
	Man	Proximal section of jejunum
	Hamster	Fatty acids are absorbed in jejunum
β-galactosidase	Rat	Maximum in jejunum
Absorption of glucose	Rat	In vitro approximately linear decrease with certain variation toward proximal portion of the intestine; also large activity in middle portion; microwaves increase transport in middle portion of the intestine
	Rat	Tumors increase transport of glucose in the first four segments of the intestine or in the entire small intestine
	Turtle	Decrease along the intestine, greatest in the first portion of the intestine
Glutamine-oxalacetic	Rat	Linear decrease by 50% along the intestine. Distribution between mucous and muscle layer
Glutamine-pyruvic transaminase	Rat	Linear decrease after 2nd section from 8
	Hamster	Linear decrease
Glycerol	Rat	After oral ingestion of glycerol-1-C^{14} only first third of the intestine contained free compounds
Goblet cells	Cat	Number of Goblet cells in crypts decreases linearly along the intestine
5-Hydroxytryptamine	Dog	Decreases linearly along the intestine
Invertase	*Oxycarenum hyalinipennis*	Greatest activity in the first ventricle but absolute content changes with diet

TABLE 3 (concluded)

Activity	Object of study	Notes
Transport of iodine	Rat	In vitro the transport of iodine across the serous toward the mucosa approaches the curve of normal distribution
Isocitric dehydrogenase	Rat	Distributed between the mucous and the muscle layer equally along the intestine
Lactic dehydrogenase	Rat	Linear decrease
	Hamster	Decrease after the second segment
Lipase	Rat	Twice higher in the first segment than in the second, then decreasing almost to 0 in the 4th segment of 8
Magnesium	Lamb	After oral ingestion the specific activity was higher in the middle section of the mucosa of the intestine than in other segments
Consumption of oxygen	Cat	Constant for the circular layer of muscles
	Other species	Change for the circular layer of the intestine
Absorption of phosphorus	Rat	In vivo in duodenum, jejunum, ileum, and colon (in %) approximately 29, 25, 38, and 8
Synthesis of salicylglucoronide	Guinea pig	Approaches the curve of normal distribution
Substrate required for ion transport	Rat	Ion transport is facilitated by the presence of citrate or pyruvate instead of glucose in the ileum of rats but not in the jejunum
Zone of the surface	Rat	Decrease away from pylorus
Excretion of tyroxine	Rat	Enters bile as the glucoronide but is hydrolyzed in the small intestine, mainly in ileum
Absorption of vitamin B_{12}	Dog	Function of ileum
Water transport	Rat	Various sections of the intestine respond differently to presence of glucose and phloridzin. Middle portion of the intestine has the greatest activity in the transport of water and glucose

distal direction. Thus, the utilization of starch is correlated more closely with the extent of membrane rather than cavital digestion.

Maximal activity of invertase is observed in the first and second quarters of the small intestine (Fig. 35), and that of peptidase in the second (dipeptidase activity was determined by the rate of hydrolysis of glycyl-L-leucine; Fig. 36). Robinson and Shaw (1960) who employed other substrates found maximal activity in more distal sections of the small intestine.

As we mentioned earlier, many contradictions are related apparently to the fact that enzymatic activity and its distribution along the intestine change depending on the functional state of the animal, and so forth.

It is important to question to what extent the distribution of enzymes along the intestine, as described above, is accidental. Apparently, there are ample conjectures to visualize the results of prolonged adaptive evolution in the spatial distribution of enzymatic activity.

The lower content of enzymes in the mucosa of distal segments is possibly a result of the fact that there the induction by substrates of the necessary level of synthesis is absent, and that these substrates are digested and absorbed in more proximal sections. Indeed, when the work of the upper segments of the small intestine is altered, the level of oligosaccharides in the distal portions markedly rises (for greater details see Chapter VI).

The idea that the low activity of distal segments is connected with the absence of substrate-enzyme induction is in good agreement with data obtained in our laboratory (Nadirova et al., 1965a). In the study of the proximal–distal gradient of the invertase of growing rats it

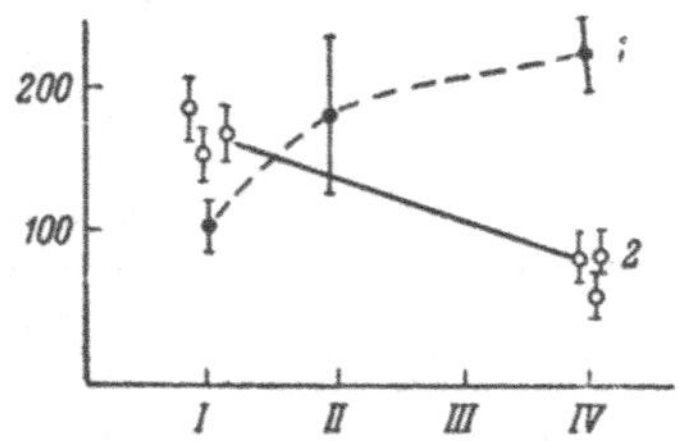

Fig. 34. Distribution of amylolytic activity between the contents of the intestine (1) and its surface (2) during digestion in rats. Abscissa: sections of the small intestine (by quarters); ordinate: amylolytic activity (in mg of hydrolyzed starch in 1 minute).

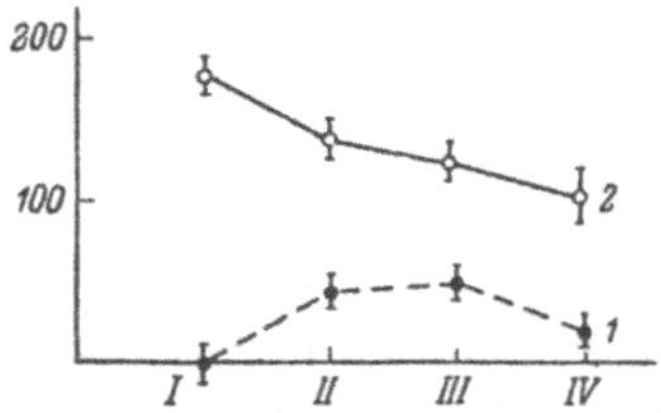

Fig. 35. Distribution of invertase activity between the contents of the intestine (1) and its surface (2) during digestion in rats. Ordinate: invertase activity (in arbitrary units). The rest of the notation is the same as in Fig. 34.

was found that the maximal activity is shifted from ileum to the first few segments of the jejunum during the 20- to 60-day period of postnatal development. During this time, the length of the intestine approximately doubles, and the ileum is located in the region which is not reached by any significant amount of the substrates. Nevertheless, two questions remain unclear: (1) why is it that the activity of invertase in the most proximal segments of the small intestine is comparatively low? and (2) why is the maximum of invertase activity markedly shifted in a proximal direction between the 20th and 30th day although the length of the small intestine during this period does not alter much?

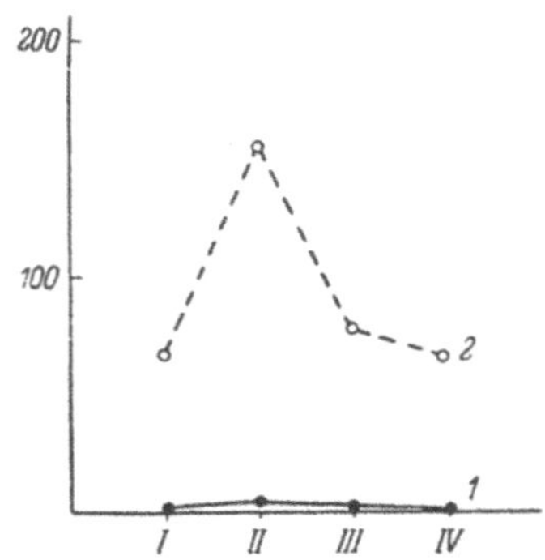

Fig. 36. Distribution of peptidase activity between the contents of the intestine (1) and its surface (2) during digestion in rats. Ordinate: peptidase activity (in arbitrary units). The rest of the notation is the same as in Fig. 34.

To this should be added the circumstance that during the entire development the intensity of membrane hydrolysis of tributyrin remains maximal in the upper segments of the small intestine (Iezuitova et al., 1964c, 1965a).

In the preceding chapter we drew attention to the fact that the transition from cavital to membrane digestion is determined not only by mixing and diffusion of the substrates, but also by the surface activity of the intermediary products of hydrolysis.

It is understandable that the lipids which possess the greatest surface activity will pass from the inside of the phase to the surface of the intestinal epithelium more intensively than the intermediary products of hydrolysis of proteins and carbohydrates. It is also clear that the surface area occupied by the lipids will be greater the higher their concentration in the nutrient ration. This is reminiscent of distribution chromatography on the porous surfaces of adsorbants.

This hypothesis makes understandable the fact that after the 20th day of postnatal life the maximum of sucrose hydrolysis is shifted in a caudal direction (during this period the percentage of milk in the ration is still high) and one observes a shift of the gradient in a proximal direction when the rats begin to eat independently (Figs. 37 and 38).

It is also understandable that in the course of the usually mixed nourishment of man lipids will be hydrolyzed in the most proximal segments of the small intestine.

Thus, the lipid components of food may make the surface of the proximal portion of the small intestine unavailable to other nutrient substrates.

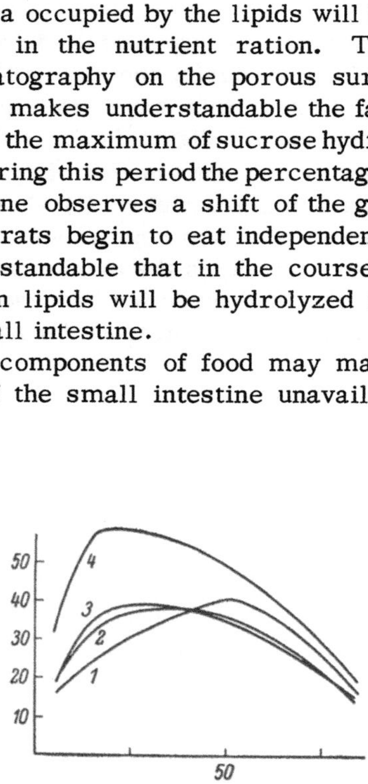

Fig. 37. Distribution of invertase activity along the small intestine of rats of various age groups. Abscissa: length of the small intestine, taken as 100%; ordinate: activity of invertase (in mg % of hexoses formed during the incubation). (1) 20, (2) 24, (3) 27, (4) 60 days of postnatal life.

One additional circumstance should be kept in mind. Apparently, the surface of the upper segments of the duodenum may be saturated with bile acids. The data pertaining to the competition between amylase and bile acids and other surface-active substances favor this possibility. Besides the proximal–distal gradient, in the course of description of the digestive and resorptive functions of the intestinal cells, one must keep in mind the unequal distribution of enzymes and microvilli in various sections of the villus.

It was indeed found (Palay and Karlin, 1959a; Brown, 1962) that the brush borders of the intestinal cells in the regions of the crypt and the top of the villi are not identical. The microvilli of crypt cells are shorter and thicker (height 0.67 micron and diameter 0.15 micron) than on the villus, particularly on its top (about 1.36 and 0.08 micron). If a single cell in the region of the crypt is 4.8 microns^2 by 1 micron^2 of the free surface of the cell, then in the region of the top it is equal to 39.2 microns^2 by 1 micron^2 of the free surface of the cell. During the investigation of alkaline phosphatase by histochemical means it was shown that its concentration is increased in the direction from the crypt to the top of the villi (Brodskii, 1962, 1964).

Brown proposed a hypothesis which could explain the increase in the number of microvilli per unit surface without the formation of new ones within each cell. This could be accomplished only if the free surface of the cell is contracting. This hypothesis, however, according to which the initially formed structures remain for the entire life of the cell, does not explain in what manner the concentration of enzymes on the membrane surface increases. From our point of view the replacement of the membrane material is possible and as a result of this, enzymes may appear on the surface of the cell. The growth of the microvilli is also not excluded. Irrespective of how the number of microvilli and the concentration of the enzymes on their surface changes, it is important to keep in mind that the cells of the tops of the villi are the most functionally active, while the least active are the cells of the crypt, as has been shown by many investigators (Patzelt, 1936; Palay and Karlin, 1959a). This intravilli gradient may be altered in pathological conditions when the cells at the top of the villi have very few microvilli and the membrane contains a very small amount of the enzymes.

STERILITY OF MEMBRANE DIGESTION

A considerable number of various bacteria, whose concentration reaches many millions of microbial bodies per 1 ml, live in the chyme

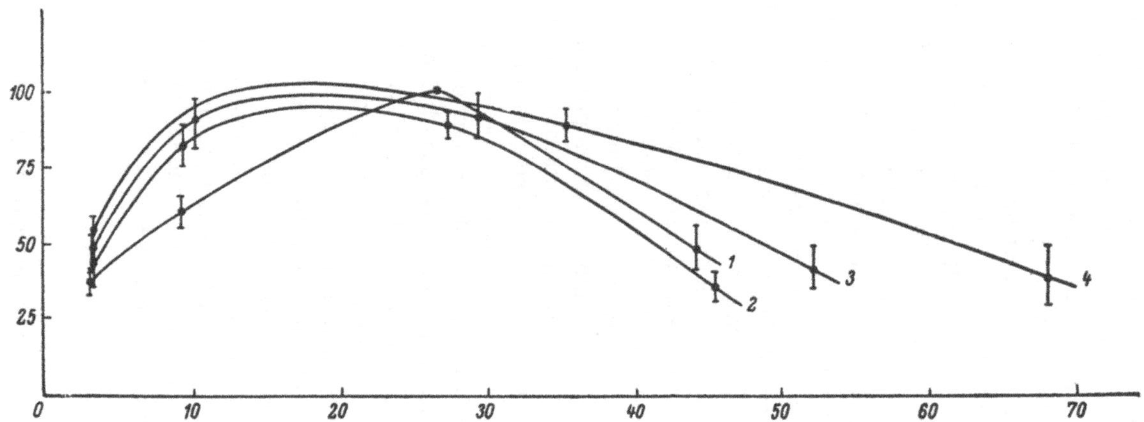

Fig. 38. Distribution of invertase activity along the small intestine of rats of various age groups. Abscissa: length of the small intestine (in cm); ordinate: activity of invertase (in % of hydrolyzed substrate during the time of incubation). For 100% was taken the maximal activity of invertase in each age group. The rest of the notation is the same as in Fig. 37.

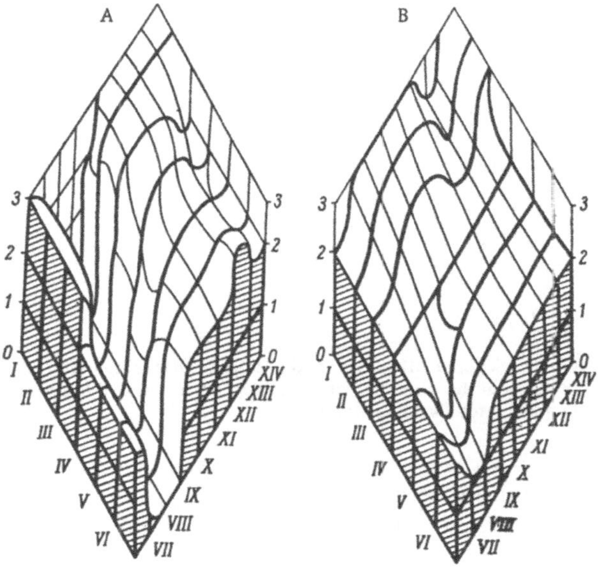

Fig. 39. Bacterial flora of the gastrointestinal tract of the rat (according to Luckey, 1964, p. 230, Figs. 1 and 2). (A) lactobaccilli; (B) gram-negative microorganisms. Abscissa: (I-VI) sections of the digestive tract: (I) secum, (II) lower portion of ileum, (III) upper portion of ileum, (IV) jejunum, (V) duodenum, (VI) stomach; (VII-XIV) various diets: (VIII) 2-day starvation, (VIII) 5-day starvation, (IX) meat, (X) meat + lard, (XI) meat + lactose, (XII) mixed diet, (XIII) diet with high fat content, (XIV) diet with high protein content; ordinate: decimal logarithm of the number of bacteria per 1 mm^3 of chyme.

of the small intestine (Fig. 39). It is well known at present that bacterial flora of the intestinal canal is useful under ordinary conditions, being a supplier of many indispensable amino acids and of certain vitamins essential to macroorganisms. A question arises, however, why this rich bacterial flora does not become a competitor to the macroorganism in the utilization of amino acids, glucose, and other nutrients.

We have drawn attention to certain circumstances which could clarify this question (Ugolev, 1961a, 1963). The final stages of digestion, in the course of which monomers are formed and are available to the microorganisms with their diffusional nourishment, occur in the ultrapores of the brush border. If one compares the sizes of the bacteria populating the intestinal canal (several microns) and the pore sizes of the brush border (100-200 Å) then it becomes clear that one of the important functions of the brush border is that of a peculiar

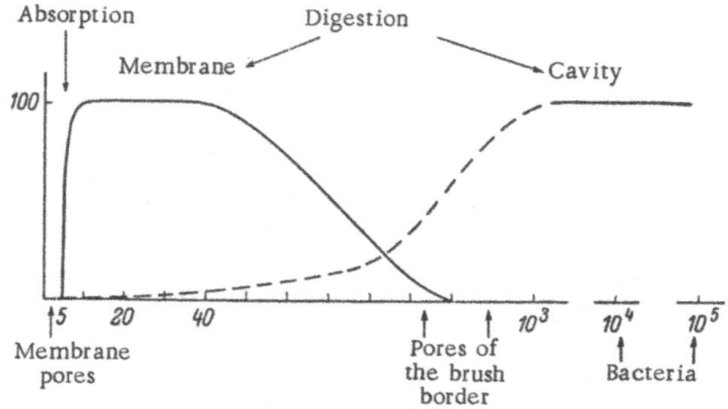

Fig. 40. Scheme illustrating the interrelationships of basic processes in the small intestine in relation to sizes of hydrolyzable substrates (water-soluble).

bacterial filter because of which the final stages of hydrolysis take place under absolutely sterile conditions (Fig. 40).

Thus, under normal conditions the loss of nutrients by the macroorganism is prevented by the fact that the final stages of hydrolysis occur in the pores of the brush border which are inaccessible to the microflora. Apparently, this circumstance is of importance for the limitation of the potentially unlimited bacterial multiplication in the small intestine.

As is shown in Chapter V, the alteration of membrane digestion and the formation of the final products in the intestinal cavity may be reasons for the rapid development of its bacterial flora.

PERISTALSIS OF THE INTESTINAL TRACT AND MEMBRANE DIGESTION

Dahlqvist and Borgström (1961), while investigating the cleavage of polysaccharides in the small intestine of man, came to the conclusion that membrane digestion does not affect the course of the initial stages of hydrolysis of soluble starch. Ugolev (1961b) obtained the opposite results in rats. De Laey and Iezuitova (1962) found that in certain of their experiments the hydrolysis of starch in the small intestine of white rats was accomplished entirely by the enzymes of the cavity, and in some other experiments mainly by the enzymes adsorbed on the surface of the small intestine. Later on Iezuitova et al. (1963a), De Laey (1963), and then Iezuitova et al. (1964a) were independently confronted by facts which resolved these contradictions. It turned out that membranal effects during low rates of diffusion are quite weak, but they increase with the increase in the rate of diffusion (Fig. 41).

Iezuitova et al. (1963a) showed that the intensity of membrane digestion of sucrose is proportional within wide ranges to the rate of diffusion. Ugolev et al. (1964b) observed that hydrolysis of sucrose as well as of dipeptides on the membrane surface is greater the more intense the movement of fluid in relation to the surface of the mucosa.

These facts are well known in regard to catalysis on porous surfaces of nonbiological origin (Germain, 1961).

The rate of a heterogeneous reaction is basically determined by three components: rate of entry of the initial products from the inside of the phase to the surface, rate of surface reaction, and the rate of removal of the products of the reaction from the surface to the inside of the phase or into the lining. If the catalytic reaction on the surface proceeds rapidly, then the limiting factor becomes the rate of entry of the substance from the phase to the surface. It is therefore understandable that improvement in the conditions of contact of the hydrolyzable substrates (polysaccharides, oligosaccharides, peptides, etc.) with the surface of the small intestine must be accompanied by an increase in the effects of membrane digestion. From this standpoint the contradictions about which we spoke at the beginning of the section are understandable.

Indeed, it is well known that with low rates of movement of fluid in a pipe (in our case, the intestine) the flow will become laminar, but near the surface a stationary layer is formed. With the increase of perfusion the flow becomes turbulent and therefore the conditions of contact between the soluble substrates and the surface of the small intestine improve. On the basis of these data a very important con-

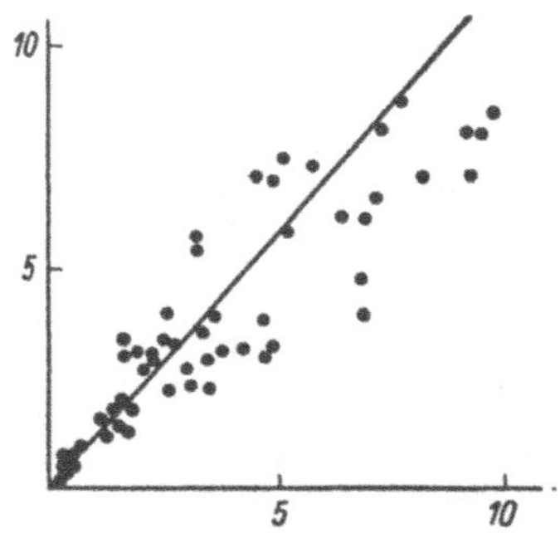

Fig. 41. The dependence of hydrolysis of starch on the rate of diffusion. Abscissa: volume outlay (in ml/minute) to capacity of the intestine (in ml); ordinate: relationship of hydrolysis in vivo to hydrolysis in vitro.

clusion can be drawn, namely, that the effectiveness of membrane digestion is determined not only by the state of the enzymatic layer on the surface of the intestine, but also by the conditions of the passage of the hydrolyzable substrates from the chyme to the surface.

In this connection an assumption has been expressed that the normal peristalsis of the intestine plays an extremely important role in the maintenance of the high rates of membrane digestion (Iezuitova et al., 1963a; Ugolev, 1963; see reviews by Spencer, 1960; Bogatch, 1961; Bülbring, 1961; Connel, 1961; Rowlands, 1961; Menguy, 1964).

On the other hand, the experiments cited above permit the conclusion that the slowdown in mixing movements of the small intestine may lead to a reduction in the tempo of membrane digestion even if the enzymatic layer on the surface of the intestine is fully intact.

ENZYMATIC LAYER OF THE INTESTINAL SURFACE

In this section we attempt to systematize the information pertaining to the enzymes which ensure membrane digestion and which we have already mentioned in various chapters of this book. We will not return to numerous facts which permit the conclusion that the final stages of digestion take place mainly, or exclusively, on the outer surface of the cellular membranes, and not intracellularly (see Chapter II).

At present it is established that disaccharidases, peptidases, as well as, apparently, nucleotidases, nucleosidases, alkaline phosphatase, and monoglyceridase, are firmly bound to membrane structures and do not pass into the cavity of the small intestine. In contrast to this group of intestine-proper enzymes, the adsorbed enzymes are redistributed more or less easily between the cell surface and the surrounding fluid. Thus, an equilibrium always exists between the enzymatic activity of the intestinal content and the intestinal epithelium with respect to these enzymes. To this group of enzymes belong pancreatic amylase, proteases, and lipase. The question of the adsorption of ribonuclease and deoxyribonuclease which are secreted with the pancreatic juice has not been studied.

As a result, an enzymatic layer is formed on the surface of the intestine which is of double origin with a very wide spectrum of activity toward the nutrients, a spectrum which is limited, as was mentioned earlier, by the pore size of the brush border and the accessibility of the hydrolyzable bonds to the action of the enzyme fixed on the surface.

The general scheme of digestion in higher animals can be reduced

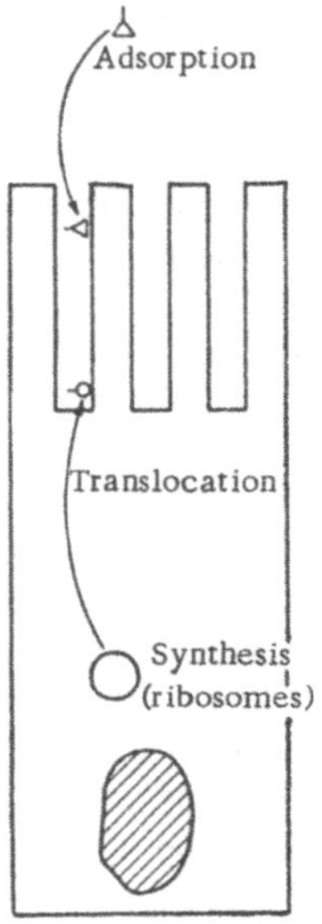

Fig. 42. Mechanism of formation of the enzymatic layer of the intestinal surface as a result of adsorption of enzymes from the chyme and translocation of the intestine-proper enzymes on the cell surface.

to the following. The initial hydrolysis takes place in the cavity of the small intestine by the enzymes dissolved in the water phase. As a result of mixing movements of the intestinal musculature, the intermediary products of hydrolysis come in contact with the surface of the villi, and, since the latter possess surface activity, the hydrolysis products enter the brush border if their sizes are not too great. This passage from the cavity of the small intestine to the brush border can also be facilitated by transfer in the current of the solvent which arises as a result of absorption of water by the intestinal cells.

We know what takes place in the pores of the brush border only by the end result. Here the oligomers are cleaved to monomers at a great rate and then are actively transported across the intestinal barrier into the inner medium.

We were not successful in our attempts to follow up the hydrolysis of dextrins (Ugolev and Marauska, 1964) and peptides (Kushak, 1965) in the brush border. In both cases we observed a rapid fall in the concentration of the initial product and the formation of free amino acids in the case of tetrapeptide and glucose in the case of dextrins.

We earlier expressed the opinion that at the base of these events a conveyor mechanism in which the intermediary products are subjected to stepwise cleavage on the surface due to spatial arrangement of the enzymes localized on the surface may be involved. Also, it may be assumed that the pores of the brush border are a peculiar "trap" for substances which penetrate them, and the probability of escape from it for the uncleaved substrates is very small (Ugolev, 1963). According to the latter hypothesis, the uninterrupted transfer of the substrate from enzyme to enzyme is not obligatory. The end effects, observed experimentally, are identical in both cases.

How does the enzymatic layer form on the surface of the intestinal cells? We have already mentioned that one should distinguish between the intestine-proper enzymes and the adsorbed enzymes of mainly pancreatic origin (Fig. 42).

The study of the desorption of proteases, lipase, and amylase (the latter has been studied in detail) showed that the passage of enzymes into the liquid phase is well described by the exponential curve. The firmness of the bond with the surface is not equal for the same enzyme in animals of different species. Thus, in guinea pigs amylase is desorbed considerably slower than in rats and mice. During alterations in functional states the rate of desorption of amylase and lipase changes. 2-4-Dinitrophenol poisoning and treatment of the intestine with acetone increase the rate of desorption of amylase and lipase. In this regard the data of Hooft et al. (1963b) that in the course of certain maladies in man the adsorption of amylase by the intestinal surface sharply falls are of great interest.

One cannot exclude the fact that the enzymes are adsorbed not at any but only at definite points of the membrane which can be considered as specific adsorption centers. In favor of this viewpoint is the circumstance that certain synthetic enzymes are practically completely not adsorbed on the intestinal surface (Iezuitova et al., 1964a).

It was shown earlier (Iezuitova et al., 1964a) that the enzymes are adsorbed from the intestinal contents (Fig. 43).

Recently Rakhimov (1965), while studying the hydrolysis of starch during perfusion in an isolated section of the small intestine in dogs by the method of Tiery–Vella, observed a desorption curve typical of such experiments with amylase. Initially he observed considerable cleavage of the substrate in the course of perfusion (in vivo) which exceeded many times the activity of the enzyme in the perfusate (in vitro). Later the activity in vitro and in vivo was about the same. After the cessation of perfusion the activity in vivo was restored, although the sole source of the enzyme under the experimental conditions in the isolated loop was blood amylase. It is possible that the adsorption of amylase excreted from the blood has a definite significance in pathological states.

De Laey in 1964 observed a considerable increase in amylase adsorbed on the surface of rat intestine after feeding as compared to

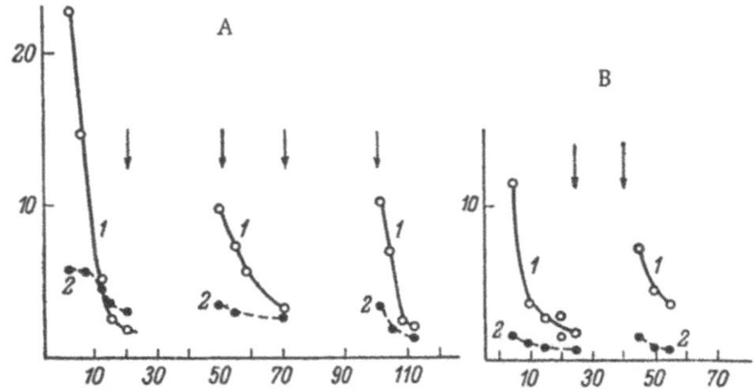

Fig. 43. Changes in amylolytic activity during perfusion of rat duodenum (entry of pancreatic juice into the intestine is preserved) according to the data of De Laey (A) and ours (B). Determination after 3 minutes by the modified method of Smith–Roy. During interruptions (arrows) perfusion was stopped. Abscissa: time (in minutes); ordinate: amylolytic activity (in mg of hydrolyzed starch/ml/minute); (1) in vivo and (2) in vitro.

starvation. Most probably this is related to a change in the concentration of amylase in the intestinal content after feeding.

Thus, one can assume a supplementary "charge" of the enzymatic layer, and from this point of view one can understand an important phenomenon described by Pavlov (1897). We have in mind here a short latent period of secretion of the pancreatic gland which becomes active long before the nutrients enter the small intestine. It could be that this advance secretion, whose significance heretofore was not clear, ensures the accumulation on the surface of the small intestine of enzymes which are essential for the processes of membrane digestion. Certainly, the increased adsorption of pancreatic enzymes on the intestinal surface depends not only on the increase in their concentration in the chyme but also on the changes in the adsorption properties of the surface. We discuss this in the section "Regulation of Membrane Digestion."

The question of the mechanism of entry of the intestine-proper enzymes on the outer surface of the membranes is not clear. Brodskii (1962) and many other authors, however, have shown that alkaline phosphatase is absent in crypt cells and appears after their migration to the zone of the villi. Thus, the membrane of the newly formed cells possesses different enzymatic properties than the membrane of the villi cells which are directly involved in digestion.

This allows the important conclusion that the enzymes of the intestine proper may appear on the surface at a time when the membrane exists. This is indirectly supported by the fact that with the

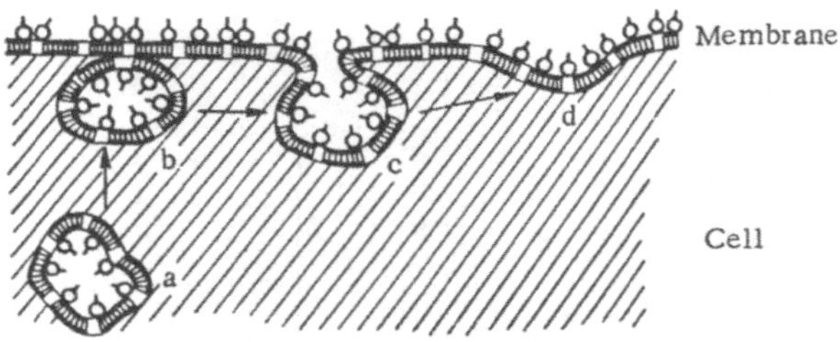

Fig. 44. Possible mechanism of transfer of the enzymes of the intestine proper to the cell surface via reverse pinocytosis. (a)–(d) Various stages of the process.

maturation of intestinal cells and their movement from the crypt to the villi the number and sizes of microvilli increases (Brown, 1962). It is difficult to imagine that the enzymes synthesized in the ribosomes in an isolated state are transported across the membrane and then are fixed on it. It is more logical to assume that the synthesis of the fragments of the membrane together with their enzymatic apparatus occurs inside the cells. These fragments are then included in toto into the composition of the cell surface. In other words, one can assume an intensive exchange of membranal blocks between cytoplasm and the cell surface. We assume that at the base of this exchange may be the mechanism of pinocytosis as a result of which a portion of the membrane enters the cell, and of reverse pinocytosis as a result of which the newly formed block enters the composition of the membrane (Fig. 44). In any case it is improbable that the mechanism of pinocytosis could play an important role in the nutrition of higher animals. On the other hand, it is adequate to ensure the reconstruction of the membrane and replacement of its injured parts.

Certain data concerning the ultrastructure and enzymatic properties of intestinal cells during irradiation damage indicate that this process of exchange of membranal substances may proceed quite rapidly, that the processes of synthesis of an enzyme and its translocation on the surface are not identical, and, finally, that these processes are interconnected (these observations are considered in the proper section of Chapter V).

In order to understand the biological balance between the processes of cavital and membrane digestion, one must evaluate not only the amount and the properties of the enzymes which are involved in the processes of the hydrolysis of nutrients, as is generally done, but also the time of useful work (useful life) of a given enzyme molecule.

During the description of cavital digestion and, therefore, of the enzymes involved therein, one must keep in mind that the enzymes of the chyme move from the duodenum in the direction of the large intestine and, therefore, their useful effect is limited to a few hours. This is understandable, since after the entry of the enzymes into the large intestine they are rapidly inactivated by its abundant bacterial flora.

In reality, apparently, the useful life of an enzyme is even smaller because the hydrolysis of the bulk of the nutrients is completed in the proximal quarter of the small intestine (Borgström et al., 1957; Spencer, 1960; Wilson, 1962).

That portion of the pancreatic enzymes (amylase, lipase, trypsin, and chymotrypsin) which is adsorbed on the surface of the intestinal epithelium lives, apparently, somewhat longer. It is difficult to say, however, to what extent the length of stay of an individual molecule in the upper portions of the small intestine changes, since the adsorption of these enzymes on the surface is reversible and, therefore, it would be correct to speak not of an increase in useful life of an individual adsorbed molecule, but to consider a certain portion of the enzymes as acting in the adsorbed state. The composition of these enzymes, naturally, may change.

The useful life of the hydrolases of the intestine proper appears to be remarkably long compared to those of the cavity. Let us assume that each enzyme molecule exists on the surface as long as the intestinal cell lives. Then one must conclude that the length of the useful life of the molecule is about two days. Therefore, at least in part, the enzymes which are involved in membrane digestion are active several times longer than the enzymes of cavital digestion.

In accord with the traditional viewpoint, with each new act of digestion there is a complete renewal of enzymes involved in the digestion.

Keeping in mind the lifetime of the enzymes involved in membrane digestion, one may introduce substantial changes in this viewpoint. Certain enzymes participate in nutritional activity for at least one to two days.

One must note, however, that at the present time we do not know the actual lengths of stay of the enzymes of the intestine proper on the surface, and this problem will be somewhat clearer after we investigate the laws governing the exchange of membranal substances between the cell content and its wall.

MEMBRANE DIGESTION AND ABSORPTION

This problem will be considered from several angles, each of which requires special methods of investigation and describes various aspects of the processes taking place on the surface of intestinal cells.

First of all, we must consider the interrelationship of the hydrolytic and transport processes in the cell membrane, which have been considered in sufficient detail in Chapter II in connection with the discussion of the localization of the final stages of hydrolysis. On the basis of various approaches we came to the conclusion that the hydrolysis of dipeptides and disaccharides occurs exclusively or mainly on the outer surface and not after these oligomers penetrate the cell membrane.

Let us now examine the same problem on the basis of data which relate to the transport of monomers and dimers. Even if it is taken into account that the data of Lindemann and Solomon (1962) and other authors that the disaccharides and dipeptides similar to their size cannot penetrate the cell membranes are not quite exact, even if we assume that modern methods do not permit the measurement of the radius of the pores for water-soluble substances, if we assume that the pores are much larger than they are generally considered to be at the present time, then still the conclusion is inevitable that the transport of dimers across the membrane will proceed considerably slower than the transport of an equivalent solution of monomers. This is determined by at least two factors: by the rate of diffusion across the membrane which is generally inversely proportional to molecular volume, and by the fact that dimers are not actively transported across the membranes in contrast to monomers. What then is really taking place? In experiments of various authors (Wilson and Vincent, 1955; Chain et al., 1960; Dahlqvist and Thomson, 1963a; Parsons and Prichard, 1965) it was found that on the introduction of disaccharides into the intestinal cavity the latter were absorbed at the same rate as the equivalent mixtures of monosaccharides.

In Chapter II we mentioned that cleavage of disaccharides takes place in the course of absorption. In order to examine this question in greater detail, special work was performed in our laboratory (Iezuitova et al., 1965a) in which was investigated the rate of absorption of several carbohydrates—monosaccharides, disaccharides, and polysaccharides. Some of these do not require hydrolysis (monosaccharides); others are hydrolyzable by the intestinal cells (disaccharides), while starch is subject to all stages of treatment, first in the cavity and then in the region of the intestinal surface. These experiments were conducted on white rats of the Wistar strain in the upper segment of the duodenum in which equivalent amounts of mono- and disaccharides were introduced. Blood samples were taken 15 minutes later by means of a polyethylene catheter introduced into the hip artery. An increase in blood hexoses compared to the initial level was measured (Fig. 45). Preliminary experiments showed that the blood-sugar level is a function of absorption and deposition of sugar. Under the conditions of the experiments, the 15-minute interval after the introduction of the substrates characterizes best the rate of absorption and does not reflect the deposition of blood sugar.

Several observations demand attention. First of all, the increase in the content of hexoses in the blood is, in general, determined by the rate of hydrolysis of disaccharides. Indeed, introduction of lactose induces considerably lower hyperglycemia (25.7 ± 8.0) than maltose (76.1 ± 4.8) or sucrose (89.9 ± 5.8). Cellobiose had no significant effect on the blood-sugar level (2.6 ± 7.0). It is of interest that the extent of the increase in the blood-sugar level on introduction of maltose and glucose was practically the same (76.1 ± 4.8 and 75.4 ± 9.2, respectively). Is this possible if the hydrolysis is accomplished after maltose passively overcomes the cell membrane of the intestinal cells? It does not appear to be because maltose is transported passively and the rate of its entry inside the cells cannot be higher than the rate of entry of actively transported hexoses, that is, it must be several times slower than the rate of resorption of glucose. In addition, since the molecular weight of maltose is greater, its diffusion must proceed even slower. And yet, as was mentioned, the hexose level in the blood after the introduction of maltose rises at the same rate as when an equivalent solution of glucose was introduced into the intestinal cavity. Even more surprising is the fact that the hyperglycemia after the introduction of sucrose is higher than after the introduction of an equivalent mixture of glucose and fructose (89.9 ± 5.8 and 51.0 ± 10.6,

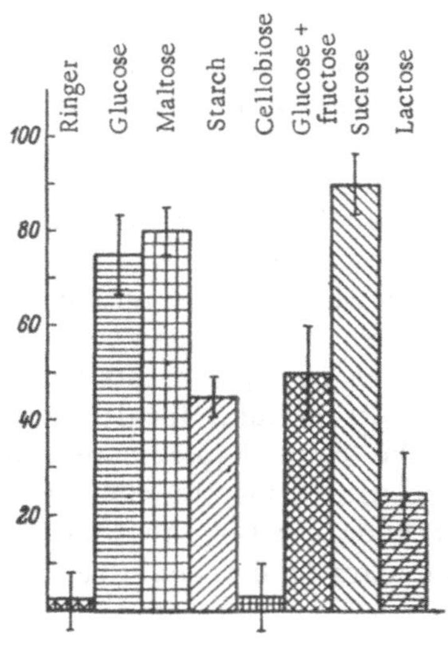

Fig. 45. Concentration of hexoses (in mg %) in blood of white rats 15 minutes after the administration into the small intestine of various mono-, di-, and polysaccharides (equivalent to 4 ml of a 5% solution of glucose).

respectively). From this one can draw only one conclusion. Transport across the membrane of the brush border is actively accomplished after the glucose formation is completed by the enzymatic hydrolysis of maltose.

It must be emphasized once more that digestion on the outer surface of intestinal membranes is a mechanism which ensures the effectiveness of the entire chain of transport processes from the intestine into the inner medium. This question was considered in detail by many investigators with respect to amino acid transport (see reviews: Wilson, 1962; Shishova-Kasatochkina, 1964a, 1964b) and transport monosaccharides (see Ponz, 1963/1964; Hart and Nissim, 1964; Nissim, 1964; as well as the review by Csáky, 1964). Csáky and Fernald (1961) conducted elegant experiments for the studies of absorption of 3-o-methylglucose (a nonmetabolizable but actively transportable analogue of natural glucose) labeled with C^{14} and came to the following conclusion: a cell membrane turned toward the intestinal cavity is relatively impermeable to sugars, and for this reason it is here where sugar pumping takes place. An accelerating mechanism located elsewhere in the intestinal wall will be useless if the first stage of resorption proves to be ineffective. It is of interest to compare these data with the hypothesis of Crane (1962) which explains the distribution of hydrolytic and transport mechanisms in the intestinal cell. According to this author, the membrane is overcome passively by the disaccharides; then in the vicinity of the inner surface of the membrane monomers are formed under the influence of hydrolytic enzymes, and only then does active transport become possible.

It is not that the authors of this hypothesis do not understand the contradictions to which such a hypothesis gives rise. Indeed, it is well known that the rate of a multistage reaction is determined by the rate of the slowest reaction, and in the case of absorption by the surmounting of the diffusion resistance of the membrane by the substances. Thus, the active mechanisms will be most effective only when they ensure high rates of transmembranal transports. The hypothesis of Crane for disaccharides and that of Smyth (1961) for dipeptides was an expedient and it arose as a result of experiments which led the authors to conclude that the hydrolysis of dimers occurs intracellularly.

In Chapter II it was demonstrated that the experiments themselves cannot be considered as proof of the intracellular localization of the hydrolytic processes. In addition, in the same chapter numerous data were cited to show that hydrolysis occurs on the outer surface of the cell membranes. Thus, there is no reason to believe that the mechanisms which perform the active transport in the intestinal cells are localized not in the region of the membrane but in the upper layers of the cytoplasm.*

Let us now compare the rate of resorption in the intestine of sucrose and of equivalent mixtures of glucose and fructose. In the experiments mentioned earlier (Iezuitova et al., 1965a) it was found that the rate of penetration of hexoses into the blood is certainly higher when disaccharides are introduced into the intestine than when the component monosaccharides are thus introduced. This phenomenon is quite surprising, but it is this phenomenon that leads to the conclusion that the processes of membrane digestion and transport are mechanisms which not only follow one another but closely interact. The greater rate of absorption of glucose after the introduction of sucrose leads one to assume that in this case more favorable conditions are created for the entry of the substances into the cell as compared with when monosaccharides are distributed from the intestinal cavity to the surface. For example, it can be assumed that disaccharidase is localized in the immediate vicinity of the pathways along which monosaccharides penetrate the cells; in this case it is clear that the probability of penetration of the molecule of the monosaccharide formed from disaccharide is greater than in the case when the same molecule is distributed from the inner phase to the surface of the cell. The subsequent fate is determined by the probability of its entry into the region of the pores which resorb it or into the indifferent region of the cell surface.

Thus we assume that membrane digestion is accomplished by enzymes which are localized in such a manner that the products of their activity have a greater probability of entering the canals along which the active transport across the cell membranes takes place than the substances dissolved in the phase. Figure 46 illustrates this hypothesis.

The hypothesis could be tested experimentally and this was accomplished in our laboratory. We assumed that if an enzyme which, for instance, cleaves sucrose is located near the entry into the pore, then its partial or total blocking should not diminish the rate of hydrolysis but should diminish the rate of penetration of the products of the reaction inside the cell. Should the enzyme be located near the inner outlet of the pore, blocking of the latter will inevitably lead to the diminution of the rate of hydrolysis. In addition, the exit of the products

*While this book was in press we noticed a new review by Crane (1966) in which he accepts the view developed here, namely, that hydrolysis of oligomers precedes the active transport across the membranes. Like many other investigators, he now connects the function of the transporter with the membrane which covers the brush border.

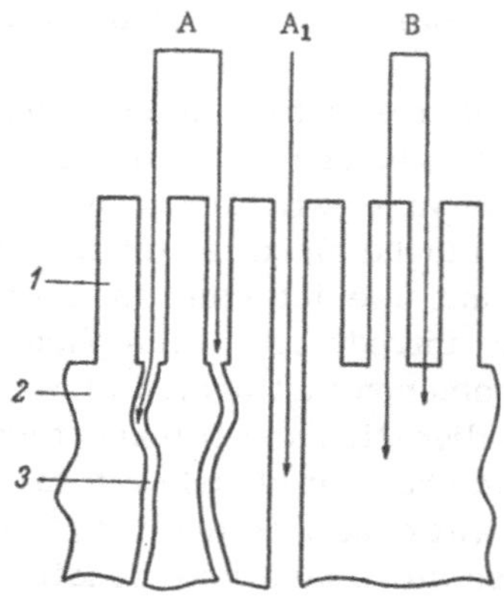

Fig. 46. Possible pathways of transport of substances in the intestine. (A, A_1) exocytoplasmic [(A) along the canals of the ergastoplasm, (A_1) along the intercellular spaces]; (B) transplasmic. (1) Microvilli; (2) cytoplasm; (3) ergastoplasmic canal.

of the reaction from the cells to the surrounding medium will be made difficult. The experiments were conducted on inverted pieces of the small intestine of white rats with the exclusion of active transport. The rate of hydrolysis of sucrose was compared with the rate of entry of the hexoses formed inside the intestinal cells. It was found that the increase in the concentration of calcium 5- to 10-fold was accompanied by a diminution in the permeability of the water-soluble pores, by an increase in the rate of hydrolysis of sucrose, and by a decrease in the rate of entry of the substances inside the cells.

It must, therefore, be concluded that the cleavage of dimers precedes the passage of substances through the pores.

At present there are no definite views regarding all the stages in all details of the transport of nutrients across the intestinal wall into the inner medium. It would appear that ample data are available for the assertion that this transport is accomplished at a high rate, that it requires energy, and that it can be accomplished against the diffusion and osmotic gradients. All this is possible thanks to the fact that the formation of actively transported substances from poly- or oligomers occurs on the outer surface of the cell membranes.

However, in order to ascertain the exact location of intestinal pumps it is necessary to describe the concentrations of substances in various sections of the intestinal cell and in the stroma of the villi. Yet the methods employed thus far failed to give the answer.

The more well-known investigations of Csáky and Fernald (1961), Csáky (1964), Miller and Crane (1961a), and Crane (1962) were devoted to studies of the distribution of hexoses in various layers of the small intestine during resorption. To this end the concentrations of substances in the fluid surrounding the mucosa from which the resorption of sugars took place, in the fluid which surrounded the serous lining into which the absorbed sugars were transported, in the scraped mucous lining, and in the muscle and serous lining together, were compared.

Thus, that which represented the mucosa fraction was in reality a total of more or less disrupted villi, including the stroma and the intestinal epithelium. Since the concentration in the mucosa fraction was higher than in the muscle–serous fraction, a conclusion was made that because of the processes of active transport the highest concentration of the substances is attained in the apical portions of the intestinal cells. Furthermore, the diffusion of the penetrated substances into the basal portions of the cells takes place passively, and from there across the basal membrane into the blood stream.

Khalimov (1963) and Ugolev (1963) developed a special histochemical technique which permitted establishing the distribution of hexoses during the hydrolysis of starch and sucrose in the intestine of white mice (in chyme as well as in the tissues) (Fig. 47). It was found that the reducing sugars in the course of digestion or after the introduction of starch and sucrose into the isolated sections of the small intestine are basically formed near the intestinal surface. A narrow line is seen in this zone which is reminiscent of the line which is revealed during the histochemical study of alkaline phosphatase and aminopeptidase. The washing out of the line occurs not toward the inside of the cell but toward the intestinal content, which indicates the localization of the process on the outer and not on the inner surface of the membrane.

Thus we see that indeed in the region of the free surface of

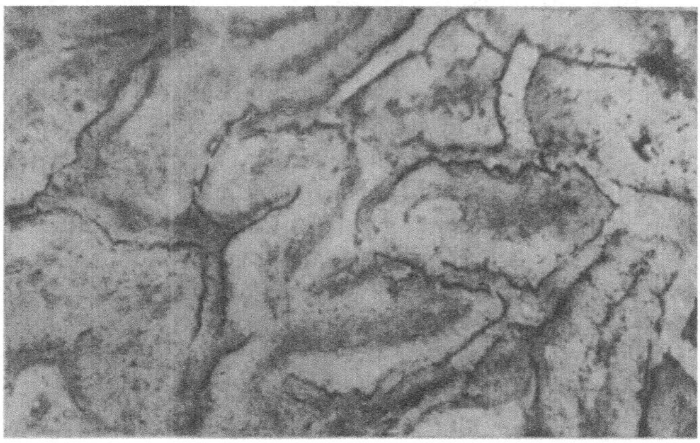

Fig. 47. Localization of reducing sugars (dark staining) in the intestine during digestion of sucrose. Preparation of intestinal section into which sucrose was introduced in advance (cross section).

intestinal cells a very high concentration of monosaccharides is reached, but what is absent is what has been supposed by Csáky (1964) and Crane (1962) and by many others, namely, there is no gradual diminution in the concentration of monosaccharides in the apical–basal direction of the cell. Examination of the preparations at higher magnifications revealed a considerable concentration of monosaccharides in the region of the brush border and in the stroma of the villi. The intestinal cells themselves gave no appreciable reaction for monosaccharides.

Thus, a positive reaction for monosaccharides is absent in the cytoplasm of the intestinal cells and it reappears in the stroma. For this reason it is not possible to detect the concentration gradient of the resorbed hexoses. The latter accumulate in the region of the brush border and in the stroma. This picture cannot be described in the frames of the accepted views and it remains for us to express two hypotheses for its explanation.

1. The active transport is accomplished along the entire intestinal cell so that the substances penetrating it leave it at a very great rate, and as a result the concentration of hexoses in the intestinal cells is considerably lower than in the region of the free surface and the stroma of the villi. In this manner two elevations of the level of concentration exist, and not one as has been previously supposed. These data do not contradict what earlier authors observed when they studied the total concentration in a fraction which consisted of intestinal cells as well as stroma of the villi but had no opportunity to localize the sugar in various microstructures of the villi.

2. The other hypothesis which can explain the active transport of monomers formed on the surface of the intestine, across the intestinal cells, is also based on the idea that the substances pass across the epithelium layer at a very great rate. It is not absolutely necessary, however, that during this process they must move through hyaloplasm, as is generally considered at present. It is possible that the active transport proceeds along special channels.

In essence such a hypothesis has not been proved and one should consider the hypothesis that the transport of monomers across the epithelial layer of the intestine may take place without penetration into the inner medium of the cell to be just as much without foundation as the one that is generally accepted. We were led to it by certain facts which are difficult to explain in any other way.

In the experiments mentioned earlier it was observed that with the increase in the concentration of calcium in sugar solutions in which pieces of the intestine were incubated, a marked increase in the rate of hydrolysis of sucrose and a decrease in the rate of penetration of hexoses into the tissues of the intestine takes place (active transport was excluded). If so, then one should expect that the end effect of this process will also be slowed down, namely, the entry of hexoses in the solution surrounding the serous lining. It turned out, however, that while the penetration of hexoses into the tissues of the intestine was slowed down, the exit of the hexoses into the serous fluid was even increased in several experiments. One can conclude from these experiments that the processes of penetration of substances inside the intestinal cells and the process of the passage of substances across the intestinal wall are not identical and to a great extent independent. Transport across the epithelial layer takes place along channels which are not blocked by calcium, in other words, not via the pores of the cellular membranes, but via special exoplasmic channels. At the present time two systems are known which could ensure such a process. These are the channels of ergastoplasm (intracellular channels) and intercellular spaces. It is important to note that many authors tend to view the channels of ergastoplasm as a system which ensures the directed transport of substances from some cellular regions to others. It appears to us possible to develop this idea by allowing the participation of ergastoplasm in transcellular transport. A series of important data in favor of the transport of water along the intercellular spaces in kidney canaliculi was furnished by Ginetsinskii (1963).

The exoplasmic hypothesis of absorption from the intestine is, apparently, adequate to explain certain important circumstances. Indeed, more and more data are accumulating at the present time to show that intracellular homeostasis is no less important a mechanism than the maintenance of the constancy of composition of the inner medium. In addition, there is ample ground to believe that, in the final analysis, the constancy of the inner medium is one of the most effective mechanisms for the maintenance of intracellular homeostasis.

The hypothesis of transplasmatic transport is tacitly based on the assumption of the completely improbable capacity of the intestinal cells to exist during changes in the concentration of sugars by several-fold, perhaps even by several tens and tens-fold. It is even more difficult to imagine that these cells could retain the ability to survive during the absorption of individual amino acids and not their equilibrated mixtures which, according to all other indications, very unfavorably affect the course of most important metabolic processes.

In essence the hypothesis of the plasmic nature of absorption must attach to intestinal cells unique properties, particularly the capacity to function normally during significant deviations of a series of constants from a normal level. Exoplasmic transport allows that the cells of the intestine, like other cells, are functioning under sufficiently stable conditions, while the transport of a huge amount of substances is performed along special channels which could pass across the cell itself or along the intercellular spaces.

In addition to the experiments with calcium, the results obtained by Iezuitova, Timofeeva, and the author, along with the experiments conducted by Kirse and the author, support the exoplasmic hypothesis. The first work was described in detail in Chapter II and in the present chapter when the role of the movement of the intestine in membrane digestion was considered (pages 98-99). These data indicate that the movement of solutions containing sucrose or dipeptides sharply increase the rate of membrane digestion. On the other hand, in the same experiments it was shown that the rate of penetration of hexoses inside the cells is not increased with an increase in convection of the fluid.

In the second work the same question was investigated except that inverted sacs and not pieces of intestine were used and the rate of passive transfer of substances inside the sac was determined. It was found that the movement of the fluid produces a sharp increase in the transfer of hexoses from the mucous to the serous side which was, apparently, in contradiction to previously obtained data. Then experiments were performed in which inverted sacs were employed and the penetration of substances into the tissues of the intestine as well as their transfer into the serous fluid was investigated. It was found that the data of both series of experiments were regularly reproducible in the same experiment. Thus, we once again are confronted with the fact that the accumulation of sugars in the intestinal cells and their transfer across the intestinal barrier are independent events and are, therefore, different processes.

True, these experiments do not permit the conclusion regarding the existence of an exoplasmic pathway, but they could be considered as a phenomenon which supports the idea expressed on the basis of the results obtained in the experiments which demonstrated the decrease in the permeability of cells under the influence of calcium. From this viewpoint the histochemical data cited in the beginning of this section become understandable. Indeed, if the transfer of substances does not take place along the entire volume of the cytoplasm but along the intercellular spaces or along the intracellular channels comprising an insignificant volume compared to that of the cell, then it becomes clear why under the microscope the cells appear to be free of sugar and why high concentrations of sugar are detected in the stroma which is a vascular–lymphatic net.

Should the assumption of exoplasmic transfer of substances across the intestinal cells be confirmed in the future, then the intracellular digestion will be seen as a mechanism which is situated away from the pathways along which the nutrient substrates move. In addition, it is possible that there exist at least two or perhaps even three types of transport among which only one is concerned with the penetration of substances across the cell membranes. It ensures the nutritive requirements of the cells, whereas transitory transport is accomplished along different channels. Perhaps, in the intestinal cells there exists such a specialization that the substances penetrating from the side of the base enter the cytoplasm of the cells, and the substances entering from the side of the intestinal cavity move exoplasmically.

The experiments with calcium distracted us from the problem of membrane digestion and absorption. We wish to mention only the simplest result of these experiments. The rate of intracellular hydrolysis limits not the transfer of dimers across the cell membranes, but the hydrolysis on the outer surface of the membranes which ensures the appearance of monomers which can then be actively transported. Because the final stages of hydrolysis are accomplished on the surface of the brush border, it is here that the high concentrations of organic monomers arise and, therefore, very favorable conditions for their resorption prevail.

Of no lesser importance is the circumstance that the processes which complete the digestion and absorption become united spatially.* It is quite understandable that neither cavital nor intracellular digestion can ensure such a high rate of absorption of the final products of digestion as membrane digestion can. We drew attention to the fact that during the introduction of dimers the rate of resorption is the same or even higher than the rate of absorption of the equivalent mixture of the monomers comprising the dimer.

REGULATION OF MEMBRANE DIGESTION

One of the greatest accomplishments of the 19th century in the field of physiology is the work of Pavlov and his school regarding the

*The close relationship of the processes of membrane digestion and active transport is supported by the data of Semenza et al. (1964) who showed that sodium ions activate the intestinal invertase of man (without change in K_m) and not only the active transport of glucose as has been established previously. The authors proposed that the site of action of the sodium ions on invertase may be considered as an entry of the glucose–sodium pump. Unfortunately, without sufficient grounds the authors were basing their proposals on the model of Crane and assumed that both effects of the sodium ions are localized under the membrane. The incorrectness of this model has been demonstrated first by us and then by other investigators, including Crane (1966). However, the basic idea of the close relationship between membrane digestion and active transport appears to us very important for the description of the interrelationships of these processes.

regulation of activity of the major digestive glands. The past decades not only confirmed the correctness of the views of Pavlov but also made the problem of regulation of the physiological functions one of the most important.

By a strange contrast we know almost nothing about how the work of another powerful mechanism performing the transformation of nutritive substances (membrane digestion) is controlled. We speak here about direct investigations of the influence of various factors on the course of membrane digestion. However, if one considers that oligosaccharidases and peptidases accomplish membrane digestion, as was shown in Chapter II, then the data of numerous investigations could be of use for the understanding of the regulation of membrane digestion, even though these investigations were conducted for different reasons.

As can be seen from Table 4, under the influence of changes in the composition of food, the content of various enzymes in the intestinal cells alters, including the enzymes which accomplish membrane digestion. The stimulating effect of the hormones of the adrenal, hypophysis, and thyroid glands on the synthesis of intestinal enzymes attracts attention.

The mechanism of the effect of hormones on the content of the intestinal enzymes is not understood in all cases. It has been shown that corticosterone acts as an inductor of the synthesis of invertase but does not affect lactase. In elegant experiments on newborn animals, Doell and Kretchmer (1964) have found that the administration of corticosterone leads to the appearance, after 2-3 days, of invertase activity which is generally absent in animals of this age. At the same time the concentration of invertase reaches a level which is characteristic of the natural induction taking place during the transition from a milk to a vegetable regimen. It is of interest to compare these data with the observations of Iezuitova, Timofeeva, and Ugolev, who, using the same doses of corticosterone, did not observe an increase in the levels of peptidase and invertase in adult rats. In connection with this the author should like to express a rather risky proposition that the induced synthesis is, perhaps, subject to an "all or nothing" law, a law by which the majority of biological units operate.

It will be remembered that from the point of view of the theory of information the transfer of information, according to the law of "all or nothing," can be considered as the simplest and the most reliable method of transfer with maximal freedom from hindrance on the basis of double code. Possibly, this simple and reliable system of regulation related basically to the induction of synthesis of enzymes becomes more complicated later on with the aid of other mechanisms of regulation. The fact is that we speak here about the synthesis of enzymes, and that it conforms with the time of appearance of the effect which coincides with the period of complete renewal of the epithelium of the villi.

We have already mentioned that the hormones of the adrenal cortex stimulate the synthesis of various enzymes which accomplish cavital as well as membrane digestion. However, the systems which synthesize various enzymes (for instance, intestinal) are not, apparently, equally sensitive to the action of corticoids. We mentioned previously that the administration of hydrocortisone induces premature synthesis of invertase in young rats but does not affect the synthesis of β-galactosidase (Doell and Kretchmer, 1964).

Levin et al. (1965) observed that after uncompensated adrenalectomy a significant fall in peptidase activity takes place, calculated for the entire intestine (drop by 50%), or when calculated per gram of wet weight of the intestine (by about 30%). Maltase activity calculated for the entire intestine drops by 23%, whereas its activity calculated per gram wet weight of the intestine does not significantly change as compared with the group of animals subjected to sham adrenalectomy. Feeding of NaCl (compensated adrenalectomy) prevents the described effects.

One should not assume, however, that under all circumstances the change in dipeptidase activity is revealed more precisely than that in case of disaccharidases. In Chapter V we present data from our laboratory which show that during irradiation damage the activity of disaccharidase changes more significantly than the activity of dipeptidase.

The distribution of enzymes along the intestinal tube may be considered as an adaptation. In a corresponding section we discussed changes in the proximal–distal gradient and the mechanism of its adaptive changes.

The regulation of biological systems includes rapid as well as slow components. The changes in the synthesis of enzymes and the changes in the distribution gradient may be considered as slow regulations. We have now the first information about the existence of rapid regulations of membrane digestion. Palay and Karlin (1959a, 1959b) discovered that in fed animals the height of microvilli is greater than in starving animals. Przheletska et al. (1961) observed that the character of the diet has a significant effect on the activity of alkaline phosphatase in the brush border. This could be considered, perhaps, as slow regulation, had not the same authors shown that soon after feeding the reaction for alkaline phosphatase becomes significantly more intense. There are data, therefore, which indicate that feeding, while affecting

TABLE 4

Effect of Food and Hormonal Factors on the Activity of Intestinal Enzymes

Influencing factor	Object of study	Changes in the activity of enzyme			Literature sources
		Decrease	Increase	Unchanged	
Protein deficiency (alimentary method)	Dog	Enterokinase, alkaline phosphatase	Invertase	Peptidases, lipase	Fomina, 1955a, 1955b
Folic acid deficiency (administration of aminopterin)	Dog	The same		Peptidases, lipase, invertase	Mikhlin, 1962
Nyacin deficiency (administration of antimetabolite, 3-acetylpyridine)	Dog	–	Enterokinase, alkaline phosphatase	The same	Mikhlin et al., 1962
Deficiency in lipotropic factors	Dog	–	Invertase	Enterokinase, alkaline phosphatase, lipase	Kadyrov, 1961a, 1961b
Deficiency in polyunsaturated fatty acids and pyridoxin	Dog	–	–	Enterokinase, alkaline phosphatase	Kuvayeva, 1962
Elevated ingestion of protein (45% casein diet)	Dog	–	Enterokinase	Alkaline phosphatase	
Vitamin C deficiency	Guinea pig	–	–	Alkaline phosphatase	Harrer and King, 1941
Zinc deficiency	Chick	Alkaline phosphatase	–	–	Roberson and Schaible, 1958
Vitamin D	Rat	–	Alkaline phosphatase	–	Crimm and Strayer, 1936
Mixed diet	Rat	–	The same	–	Jackson, 1952
Phosphoproteins	Rat	–	The same	–	Tuba and Dickie, 1955
Methionine	Rat	–	The same	–	Triantaphyllopoulos and Tuba, 1959

Tryptophane	Rat	Alkaline phosphatase	–	–	
Low fat content	Rat	–	Hexokinase	–	Long, 1953
Starvation	Rat	Amylase	–	–	Ju and Nasset, 1959
	Rat	Alkaline phosphatase	–	–	Tuba and Robinson, 1953
Na-phenylphosphate	Chick embryo	–	Alkaline phosphatase	Acid phosphatase	Kato and Moog, 1958
Adrenalectomy	Rat	–	–	Alkaline phosphatase, acid phosphatase	Tissières, 1948a, 1948b
	Rat	Alkaline phosphatase	–	–	Verzär and Sailer, 1952
	Rat	Adenosine-triphosphatase	–	Hexokinase	Lüthy and Verzär, 1954
Adrenaline	Rat	Peptidases, invertase (smaller effect)	–	–	Iezuitova et al., 1964b
Deoxycortico-sterone	Rats (9 days old)	–	Invertase	Lactase	Doell and Kretchmer, 1962
Thyroxin	Rats (9 days old)	–	Alkaline phosphatase	–	Ponz, 1945 (quoted by Spencer and Knox, 1960)
Deoxycorticosterone	Rats	–	The same	–	Halliday, 1958
Castration	Rats (males)	–	–	Alkaline phosphatase, acid phosphatase	Tissières, 1948b
Hyperthyroidism	Dog	–	Invertase	–	Fink, 1944
Growth	Guinea pig	–	Alkaline phosphatase	–	Moog and Ford, 1957
Age	Swine	–	Invertase	–	Becker et al., 1954
Colchicine	Rat	Alkaline phosphatase	–	–	Fredricsson and Wirsén, 1956

various functions of the organism, alters the ultrastructure of the brush border and the activity of at least one of the enzymes located therein.

We previously mentioned that the composition of the enzymatic layer of the intestinal surface is adapted to the quality of the nourishment. In other words, an increase or decrease of substrates in food is accompanied by an increase or decrease in the level of the corresponding enzymes in the intestinal cells. However, the mechanism of adaptation, apparently, cannot be reduced to direct inductions of the synthesis of enzymes by their substrates. The investigations of Blair et al. (1963) in particular indicate this. They observed very complicated interrelationships between the nourishment and the content of invertase activity in homogenates of the small intestine of white rats. During starvation there was a sharp decrease in the activity of this enzyme for the first 2-4 days. Diets containing large quantities (70%) of sucrose and lactose elicited a considerable increase in the level of invertase activity (compared to control diet which contained casein and no carbohydrates).

It is of interest that the increase in the invertase activity was observed on diets containing maltose or fructose and fructose plus maltose in equivalent amounts. It would appear that these experiments permit the conclusion that invertase activity is increased under the influence of all carbohydrates. However, this is not the case. It was observed that diets which contained glucose, mannose, xylose, or lactose did not increase invertase activity, while raffinose significantly decreased it. On analysis of the obtained results it is comparatively easy to explain why fructose or maltose stimulated invertase activity. In the first case we must assume that the products of hydrolysis of sucrose, particularly fructose, are the inductors. In the second case it should be kept in mind that the intestinal cell contains about five maltases of which two also possess invertase activity (Auricchio et al., 1965). If so, then it is difficult to understand why maltose increases the invertase activity while glucose, which is the product of its hydrolysis, does not. Why does not lactose stimulate this enzyme, while galactose does? In the final analysis we must come to the conclusion that the mechanism proposed by Jacob and Monod (1964) for the explanation of the induced synthesis of disaccharidases in the bacterial cell is less useful for the description of these processes in intestinal cells. It is also evident that these data indicate the existence of effects of food on the synthesis of enzymes.

Rakhimov in the Unusov laboratory investigated the hydrolysis of starch by the isolated in situ slices of small intestine of fed and starved rats. As can be seen from Fig. 48, the desorption of amylase from the intestinal surface of fed animals is considerably slower than in starved animals. These data indicate not only the alteration in membrane hydrolysis of starch under the influence of food but also the increase in the capacity of the intestine to adsorb these enzymes. As we shall see in Chapter V, under certain disease states this capacity is noticeably weakened; this leads to the inability to digest starch.

Thus, starvation substantially affects the activity of the intestine-proper enzymes (alkaline phosphatase) and also the capacity of the small intestine to adsorb pancreatic amylase on its surface. However, alteration in the activity of various enzymes involved in membrane digestion could be different, as has been shown by Levin et al. (1965). They compared the effect of 3-day starvation on the activity of maltase and peptidases which cleaved glycyl-glycine and leucyl-glycine. It was found that the peptidase activity of the entire intestine did not change compared with activity in control animals, while the activity of maltase decreased to 30%. On recalculation per gram wet weight of the intestine it was found that the activity toward glycyl-glycine increased by 61% and toward leucyl-glycine by 27%. On similar recalculations the activity of maltase in starved animals did not differ from that in the controls.

An increase in the adsorption of amylase two hours after heat stress was noted by Rakhimov (1965).

Iezuitova, Timofeeva, and Ugolev investigated the effect of adrenaline on the invertase and dipeptidase activity in white rats. To this end we compared the rate of hydrolysis of sucrose and alanyl-serine by the inverted sections of the intestine in vitro 20 minutes after the administration of a Ringer solution (control) or 1 ml of a 0.01% solution of adrenaline (experimental). The results were somewhat un-

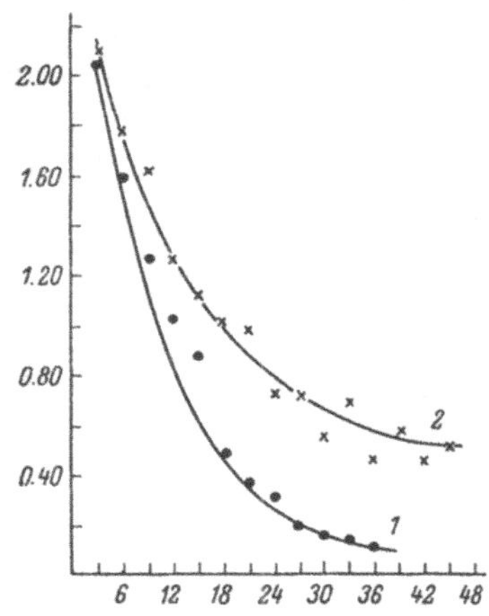

Fig. 48. Hydrolysis of starch during perfusion of the middle portion of the small intestine in starved (1) and fed (2) white rats. Abscissa: time of perfusion (in minutes); ordinate: amylolytic activity (in mg of hydrolyzed starch per minute).

expected in that we observed a sharp inhibition in peptidase activity and a relatively small but significant inhibition of invertase activity. The reason for the unequal effect of adrenaline on the cleavage of both substrates is not clear. It may be supposed the greater effect of adrenaline on the cleavage of dipeptides is determined by the change in the charge of the intestinal surface, as it is known that adrenaline acts as an agent which elicits the polarization of membranes. This hypothesis could also explain the relatively small effect of adrenaline on the nonionized molecules of sucrose.

In the section "Interrelationships of Cavital and Membrane Digestion" (page 78) we cited the experiments of Ugolev and Marauska (1964). These authors investigated the kinetics of hydrolysis of starch in the intestine by spectrophotometry of the iodine–polysaccharide complexes in a solution which was perfused through the intestine. They found that the solution was free of dextrins which show absorption in the red region of the spectrum. When, however, the perfused section of the intestine was treated with 2, 4-dinitrophenol, the membrane digestion was retained but in the hydrolysate (as during the action of amylase in vitro) there appeared a considerable quantity of erythrodextrins.

This problem requires further analysis. However, the possibility that the dissociation of respiration and oxidative phosphorylation in the intestinal cells somehow affects the conditions of transition from cavital to membrane digestion, resulting in erythrodextrins accumulating in the intestinal cavity, is not excluded.

Of course, the available data are completely inadequate for the understanding of the mechanism of regulation of membrane digestion. Yet, the available isolated pieces of information suggest that mechanisms which regulate membrane digestion exist, and that in this regulation the following indicators participate: (1) peristalsis of the gastrointestinal tract which controls the intensity of passage of nutrients from the chyme into the brush border; (2) the size of the pores of the brush border; (3) intensity of the flow of the fluid inside the intestinal cell, which in turn should be reflected in the rates of entry of the substances into the brush border; (4) composition of the enzymatic layer; (5) degree of polarization of the membrane; and (6) adsorption properties of the membrane.

There are no direct data for the above-mentioned indicators, and the nature of the factors which effect this control has not been studied adequately.

Chapter V

Membrane Digestion and the Pathology of the Gastrointestinal Tract

In one of the first communications devoted to membrane digestion (Ugolev, 1960a, 1961b) a hypothesis was expressed that insofar as this mechanism exists there must be pathological states in which the alterations in this mechanism play an important and even a decisive role.

By 1963 certain data appeared which permitted a preliminary characterization of the diseases of membrane digestion, based on numerous direct and even more numerous indirect data.

When certain general regularities characterizing the work of any mechanism are known it is not difficult to predict the basic forms of its pathology. Of course, it is much more difficult and sometimes impossible to decide which changes occur frequently and which occasionally and whether or not the alterations are sufficiently clear or are the result of certain as yet unknown factirs which will intermix with each other.

First of all, the theoretically possible forms of alterations of membrane digestion are as follows: (1) alterations in the structure of the villi; (2) alterations in the ultrastructure of the surface of the intestinal cells; (3) changes in the enzymatic layer of the intestinal surface which may or may not be coupled with changes in the ultrastructure of the brush border; (4) changes in the adsorption properties of the cell membranes; and (5) disturbances in peristalsis during which the transfer of substrates from the intestinal cavity to its surface is altered. These types of alterations, except the last one, are described below in detail.

It is appropriate to ask, at this point: why is it that the pathology of membrane digestion was noticed only recently? There is a two-way answer to this question. It is quite evident that the clinician in his thinking and working is not independent of the existing views of the theoretical sciences. It is, therefore, understandable that as long as the resorption of the products of digestion was considered to be the basic function of the intestinal epithelium, its digestive functions could not attract the attention of the physician. It was only in 1949 that major gastroenterologists such as Babkin* and Frazer (1949) with all assuredness stated that the basic and practically the only function of the intestinal epithelium was the function of absorption. It was only by the end of the 50's and the beginning of the 60's that the digestive functions of the intestinal epithelium became of obvious foremost importance, and methods were intensively developed to permit the evaluation of the functional state of intestinal cells. The first tests in this direction were quite successful and they became the basis for new views regarding the origin of certain digestive disturbances. And this was an important stimulus for further development of methods of investigation and diagnosis.

Pavlov constantly drew attention to the fact that progress in the theoretical field is most often based on the appearance of new methodological possibilities. As regards the pathology of membrane digestion this idea seems particularly justifiable. Therefore, before we begin a description of the various forms of the pathology of membrane digestion it is necessary to describe at least briefly the basic methods of investigating the digestive functions of the intestinal epithelium in man.

METHODS OF INVESTIGATING THE DIGESTIVE FUNCTIONS OF THE INTESTINAL EPITHELIUM

We will not dwell on the history of this question since it would take us too far afield. We will also not discuss the evaluation of particular advantages and disadvantages of various methods of investigation in great detail, since in this field only a very experienced clinician can escape gross errors. Let us now review the basic methodological approaches.

Aspiration biopsy is the most important, indisputable method which, most importantly, can furnish a variety of information. This method furnishes the clinician with almost all the means available to the experimenter. In essence this method is as follows. A catheter is inserted in the usual way into the gastrointestinal tract of man. The olive of the catheter is replaced by a special capsule which is capable of sucking in a piece of the surface mucosa and then slicing it. Figure 49 illustrates a capsule of modern construction. As can be seen, the capsule ensures that one obtains samples of the mucosa and slices only the mucosa without injuring the layers below it. The mucosa rapidly regenerates, and aspiration biopsy may be considered a safe

*The cited monograph of 1960 is the Russian translation of the English work published in 1949.

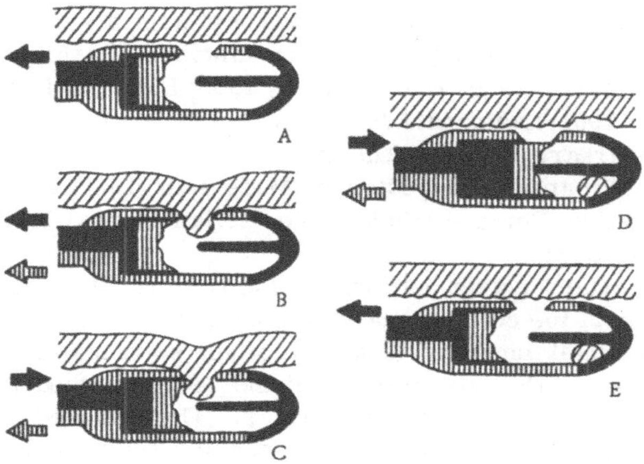

Fig. 49. Various moments of taking pieces of mucosa of the small intestine by means of aspiration biopsy (according to Ross and Moore, 1963, page 532, Fig. 3).

method at least in experienced hands.* By combining the insertion of the aspiration capsule with X-ray control it is possible to take samples of the mucosa at the desired segment of the small intestine (certain modern capsules permit taking a series of samples from various segments of the small intestine with a single insertion; in this case the sample obtained is washed out with the flow of fluid which circulates in the catheter and the capsule).

The following procedures were applied: histological and histochemical investigations of mucosa, electron microscope determination of enzymatic properties of the intestinal epithelium (usually in conjunction with thin-layer chromatography), studies of metabolic processes and active and passive transport in surviving samples and in homogenates. Aspiration biopsy makes possible, after appropriate preliminary administrations of radioactive labels and dyes, the study of various dynamic aspects, for instance, the rate of regeneration of the intestinal epithelium.

Although this method was first published about 10 years ago (Shiner, 1956a, 1956b, 1957; Doniach and Shiner, 1957; Sakula and Shiner, 1957; Crosby and Kugler, 1957), at the present time it is being widely used (Bolt et al., 1960; Flick et al., 1961; Ashworth and Chears, 1962; Kodousek et al., 1963; Plosscowe et al., 1963; Bolt, 1964; Sheehy, 1964; Madanagopalan et al., 1965; Rubin and Dobbins, 1965). There

*After the analysis of 700 biopsies, Sheehy (1964) observed no perforations and only one case of hemorrhage. He further indicated that after 2538 biopsies described in the literature only two cases of perforations and two cases of hemorrhage were observed.

are many various forms of construction of aspiration capsules; each one has definite advantages and disadvantages (Raskin et al., 1958; Brandborg et al., 1959; Hennig et al., 1959; Baker and Hughes, 1960; Flick et al., 1961; Lehmann, 1961; Ross and Moore, 1961, 1963; Crosby, 1963; Pimparkar, 1964).* The first Soviet constructions have also appeared. A capsule suggested by Masevitch (1965) and Ryss (1965) and developed by Dumesh (1965) in the Latvian medical institute should also be mentioned.

To a certain extent aspiration biopsy may be replaced by the investigation of the cellular material obtained by the usual catheterization of the small intestine (analysis of the solid matter of intestinal juice). Clearly, the possibilities of the latter method are basically limited; this is not true of the former. The removed epithelium is rapidly autolyzed and, therefore, the speculations regarding the structure of living cells (particularly the ultrastructure) may be quite limited or, possibly, inaccurate.

The data concerning the intestinal enzymes, obtained by using the former method, are of greater value, but even in this case the localization of enzymes and in certain aspects the composition will not reflect the situation in which the intestinal cell performs its resorptive and digestive functions while being a part of the villus. Nevertheless there are, apparently, a sufficiently great number of problems for which the study of the solids of the intestinal juice may prove of value. Numerous investigations published earlier testify to this (see reviews: Mikhlin, 1962; Frolkis, 1964; Shlygin, 1964).

For the characterization of digestive and resorptive functions of the intestinal epithelium, the method in which special administrations of food are coupled with subsequent determination of the products of absorption into the blood and urine is of particular importance. In essence this is not a new method, but a clever coupling of several administrations which may yield very valuable information regarding the state of the gastrointestinal tract.

Let us examine this method in detail. When glucose is introduced into the digestive tract, the resulting hyperglycemia may be considered a result of the rate of absorption and the rate of distribution of glucose (see review: Leibson, 1962). In experiments on rats we have shown that the change in blood sugar after 15 and 30 minutes (particularly after 15 minutes) is basically determined by the rate of absorption of glucose and in a considerably smaller degree by the distribution mechanisms. When starch is administered, the observed hyperglycemia is determined by the rates of the following processes: cavital

*The use of the capsule in a child clinic was described (Kauder and Bayless, 1964; Prader and Auricchio, 1965).

digestion, membrane digestion, adsorption, and distribution. The interpretation of the results obtained after the administration of starch is extremely difficult. But if one couples the administration of starch with the administration of maltose and glucose to the same individual then it is possible to make far-reaching conclusions. For instance, the absence of an increase in blood sugar after the administration of starch during typical hyperglycemia after the administration of glucose and maltose is important proof for the deficiency in amylolytic function. In other cases one may diagnose a deficiency in disaccharidases which, as has been shown earlier, is in essence a deficiency in membrane digestion. Finally, by the same method one may determine various alterations in the processes of absorption (for details see review: Holzel, 1965).

Administration of disaccharides with food permits the detection of specific enzymatic deficiencies which we discuss in one of the following sections. In an analogous manner fatty and nitrogeneous foodstuffs may be employed. These methods, however, were not as well developed and, perhaps, for purely technical reasons they were not as widely used.

The methods employed by Borgström et al. (1957), Dahlqvist and Borgström (1961), and others are of independent value. In essence they consist of the study of intestinal contents following the administration of special foodstuffs by means of multichannel catheters which permit taking samples from various segments of the intestine.

This does not exhaust all the methods of investigation of the enzymatic function of the intestinal epithelium, but these methods, especially the aspiration biopsy and the method of administration of foodstuffs, yielded the most valuable information regarding the alteration of membrane digestion.

THE ALTERATION OF STRUCTURE AND ULTRASTRUCTURE OF THE INTESTINAL SURFACE

It was mentioned earlier that the high rates of hydrolysis on the surface of the intestinal wall, the sterility of this process, and other peculiarities of membrane digestion depend on the structural peculiarities of the intestinal epithelium and, particularly, on the presence of ultraporous structures (the brush border).

From the data cited in previous sections of this book it is clear that a reduction in the digestive surface at the expense of the number of villi or microvilli per unit surface must lead to a reduction in membrane digestion.

In the past few years it has been observed that these phenomena are real and more common than had been previously supposed. The atrophy of the villi is common in a variety of diseases (Butterworth and Perez-Santiago, 1958; Laster and Ingelfinger, 1961; Trier et al., 1963; Sheehy et al., 1963, 1964; Frazer et al., 1963; MacDonald et al., 1964b; Johnston and Adams, 1964; Jeffries et al., 1964; Isselbacher and Senior, 1964; Salem and Truelove, 1965; Swanson and Thomassen, 1965). It was observed in tropical and nontropical forms of sprue, in Asiatic cholera, following neomycin therapy and various surgical interferences of the gastrointestinal tract, for instance, gastronostomy and subtotal resection of the stomach (Lundh, 1958, 1962). During

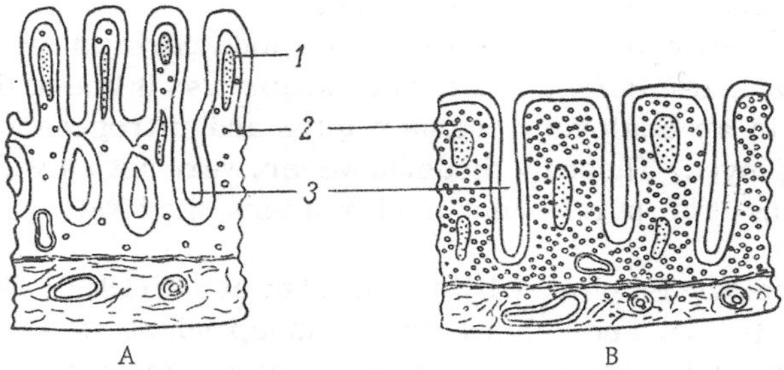

Fig. 50. Schematic illustration of changes in the mucosa of the human small intestine under normal conditions (A) and during nontropical form of sprue (B) (according to Ashworth and Chears, 1962, page 882, Fig. 3). (1) Villi; (2) lymphocytes; (3) crypts.

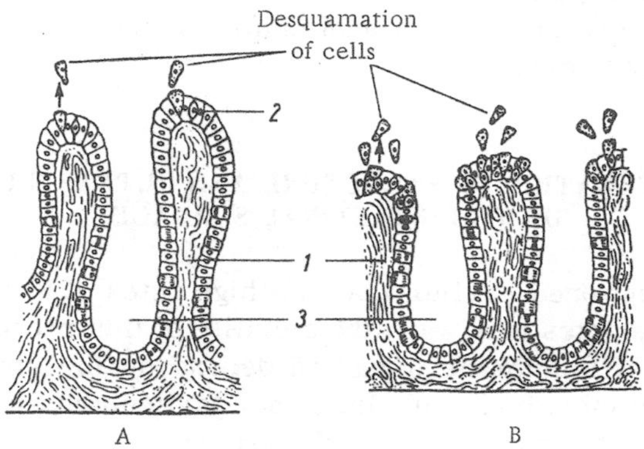

Fig. 51. Scheme of normal migration of the epithelium of crypts and villi of the intestinal mucosa (A) and increased desquamation of the epithelial cells during nontropical form of sprue (B) (according to Ashworth and Chears, 1962, page 889, Fig. 19). The notation is the same as in Fig. 50.

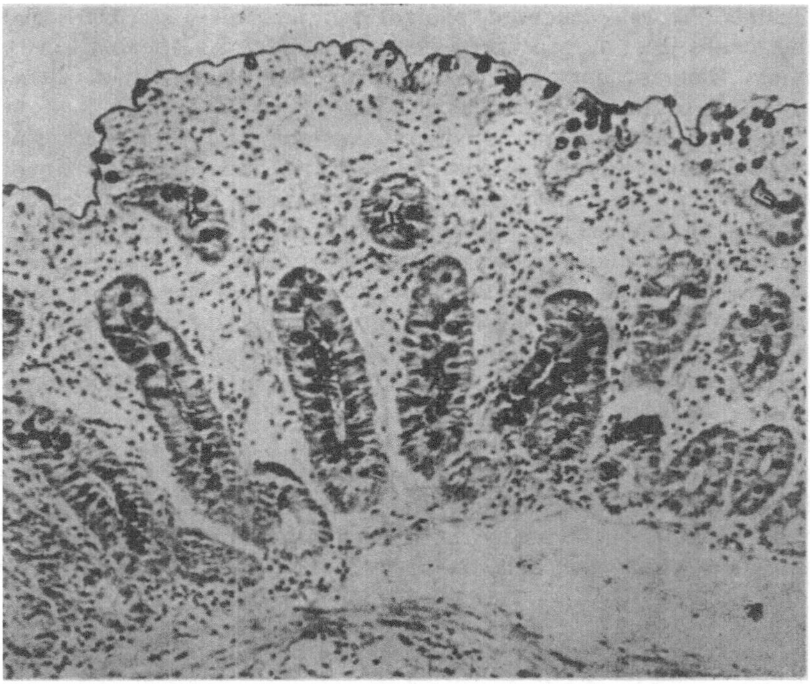

Fig. 52. Photomicrograph of the intestinal mucosa (biopsy) of a patient with the syndrome of malabsorption as a result of resection of the stomach (reduced 15% for reproduction) (according to Ashworth and Chears, 1962, page 882, Fig. 4).

partial atrophy the villi become shorter and thicker, and the crypt portion increases (Figs. 50-53).

Keeping in mind the basic objective of this book we will not discuss the pathogenesis of the atrophy of villi as well as other diseases which lead to the alteration of membrane digestion and limit ourselves only to a statement of an important fact, namely, that during the alteration of membrane digestion a wide circle of diseases occurs. It is of interest to mention that in many cases of total or partial atrophy of villi the rate of defoliation of the intestinal epithelium sharply increases; however, this is not followed by an improvement in the final stages of digestion and, therefore, provides one more proof that defoliation of the intestinal epithelium does not perform special digestive functions.

As could be expected (as we understand the mechanisms of membrane digestion), atrophy of the villi plays a smaller role in the alteration of hydrolysis of nutrients than the alteration of the ultrastructure of the brush border and the enzymatic layer of the membrane surface.

Indeed, it was observed (see review: Jeffries et al., 1964) that in many cases the therapeutic effect in celiac diseases is revealed earlier, that is, during digestive processes rather than upon the restoration of the structures of the villi.

Electron microscopic study of the intestinal epithelium during various diseases connected with the alteration in digestion and absorption led to a considerably better understanding of the causes of the diseases. In this respect the tropical form of sprue and various celiac diseases, particularly, were studied thoroughly. In these diseases serious alterations in the ultrastructure of the brush border occur, these take the form of a considerable reduction (and sometimes complete absence) in the number of microvilli and their shortening. Thus if, under normal conditions, 2000-4000 microvilli are distributed on the free surface of a single intestinal cell, in sprue or in celiac diseases their number is reduced to a few hundred, and during particularly serious forms to only a few dozen (Fig. 54). As a result not only is the effective surface of absorption reduced but in fact the porous reactor which, under normal conditions, is the surface of the intestine is destroyed. Hartman et al. (1959) were some of the first

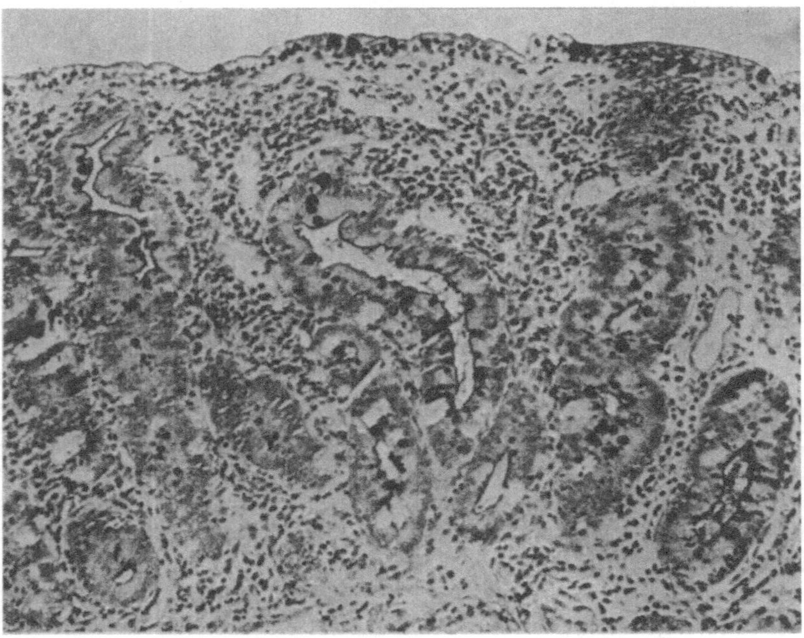

Fig. 53. Photomicrograph of the intestinal mucosa (biopsy) of a patient with a nontropical form of sprue. Obliterated villi, elongated intestinal crypts (according to Ashworth and Chears, 1962, page 881, Fig. 2).

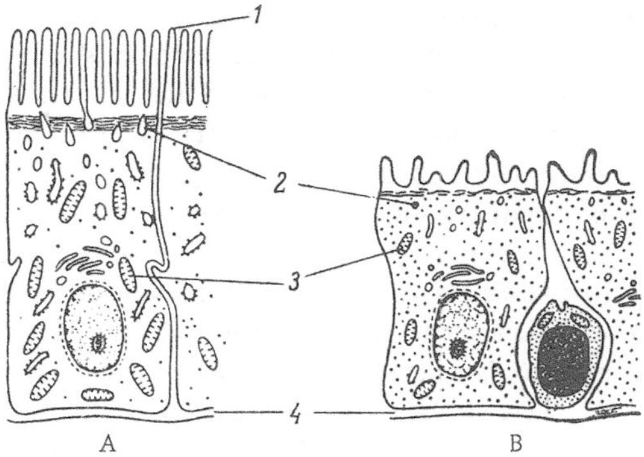

Fig. 54. Schematic illustration of the ultrastructure of cells of the intestinal epithelium under normal conditions (A) and during nontropical form of sprue (B) (according to Ashworth and Chears, 1962, page 889, Fig. 18).

who compared the brush border of normal individuals with that of individuals suffering from the nontropical sprue* by using the electron microscopic technique. In these patients, they observed a sharp decrease in the number of microvilli per unit of cellular surface. Zetterqvist and Hendrix (1960), Rubin et al. (1960, 1962a, 1962b), and Trier and Rubin (1964, 1965) confirmed these data (occasionally they observed an almost complete absence of microvilli in patients with celiac disease). In addition, these authors observed the restoration of the normal structure of the brush border after treatment of the patients with a diet which was free of gluten.

Recently similar data were obtained by several authors (Hartman et al., 1959, 1960; Sleisenger, 1961; Floch et al., 1962; Rubin et al., 1962a, 1962b; Rybak, 1962; Shiner et al., 1962; Trier, 1962; Laws et al., 1963; Samloff et al., 1965).

ALTERATION IN THE COMPOSITION AND STRUCTURE OF THE ENZYMATIC LAYER OF THE INTESTINAL SURFACE

A large number of enzymes are concentrated in the brush border; they form a powerful enzymatic layer on the surface of the intestine.

* Nontropical form of sprue is a synonym for idiopathic steatorrhea, celiac disease, and intestinal infantilism. From a purely superficial point of view, the nontropical form of sprue can be considered as one of the variants of malabsorption which is characterized by steatorrhea and alteration of absorption of various organic and inorganic substances (particularly fats). In addition to disturbances in digestion and absorption, this syndrome is characterized first of all by a sharp loss in weight.

It is possible that this layer possesses a definite organization which ensures the greatest effectiveness of the work of enzymes localized therein as well as subsequent resorption of the products of hydrolysis. It is clear that an alteration in the structure of the brush border as a result of a sharp decrease in the number of microvilli is followed by a decrease in the capacity of the enzymatic layer. Such data were supplied by a series of investigators.

However, an impression is formed that even during these coupled alterations in the ultrastructure of the brush border, enzymatic deficiency parallelism is far from being observed in every case. Padykula et al. (1961), Padykula (1962), and Jeffries et al. (1964) observed that in many patients suffering from nontropical form of sprue the ultrastructure of the brush border is preserved. At the same time the alkaline phosphatase and ATP-ase are present on the surface of the small intestine in amounts considerably smaller than under normal conditions.

These data could be regarded as evidence of the enzymatic deficiency of membrane digestion, although because of the small number of investigated indicators these data do not permit the characterization of the entire combination of alterations of the enzymatic systems of the intestinal cell. Apparently, the true picture is considerably more complex. Thus, Samloff et al. (1965), employing aspiration biopsy, showed that histochemical alterations during celial disease include deficiency in acid phosphatase, succinodehydrogenase, cytochrome oxidase, leucineaminopeptidase, DPN-diaphorase, and TPN-diaphorase. ATP-ase activity was reduced, occasionally to almost complete absence. It is of interest that after only 3-5 days of a gluten-free dietary regimen the ATP-ase activity in these patients is restored, whereas for the restoration of the activity of other enzymes this period of treatment is insufficient.

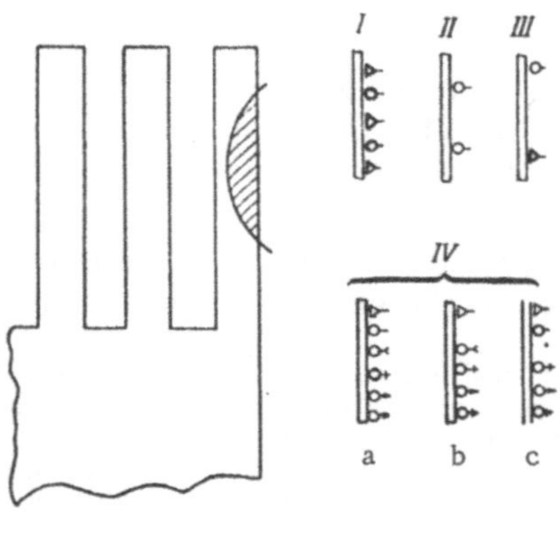

Fig. 55. Various forms of alteration of the enzymatic layer of the intestinal surface (scheme on the left). (I) Normal structure of the enzymatic layer; (II) absence of adsorption from chyme with preservation of the intestine-proper enzymes; (III) reduction in concentration of all enzymes on the surface; (IVa) normal composition of the enzymatic layer (b) and (c) absence of one of the enzymes on the surface.

It was noted that during nontropical form of sprue the content of enzymes which are involved in membrane digestion is substantially reduced as follows: alkaline phosphatase, 16.3%; invertase, 16.2%; maltase, 25.3%; isomaltase, 17.5%; lactase, 2.5%, as compared with the activity in healthy individuals (Plotkin and Isselbacher, 1964).

Large amounts of intact sucrose are absorbed in the small intestine of patients having tropical sprue. Usually sucrose is hydrolyzed by the invertase which is present in normal mucosa. The amount of this enzyme is sharply reduced in sprue (Santini et al., 1960; Rajan et al., 1961).*

Later on, various cases became known in which the structure of the brush border remained practically intact, and yet a deficiency in one or several enzymes (Fig. 55) which are normally present on the intestinal cell surface was observed. It became evident that many cases of intolerance of various types of food which remained mysterious for hundreds of years are connected with a deficiency or absence of enzymes on the intestinal surface.

It is convenient to begin the characterization of the alterations in the enzymatic layer in man by considering the well-known phenomenon which was understood only recently.

There are a large number of people who cannot tolerate milk. In some, this intolerance is revealed only after consuming large amounts of milk; in others abnormal symptoms appear after intake of minimal quantities. The extent of the disturbance varies within a wide range. In some, it is limited to a short period of disturbance in stool; in others serious forms of diarrhea, nausea, vomiting, and even convulsions arise. Adults suffering from milk intolerance were investigated (Auricchio et al., 1963; Dahlqvist et al., 1963; Kern et al., 1963; Klotz, 1964; see also reviews: Isselbacher and Senior, 1964; Jeffries et al., 1964; Dunphy et al., 1965; Catrecasas et al., 1965; Prader and Auricchio, 1965). It was found that in the majority of the cases of milk intolerance studied, the cause was simple: it was due to a sharp decrease or total absence of lactase in the intestinal cells. The absence of lactase can be established by two methods: by measurements of lactase activity in sections obtained by aspiration biopsy and by the method of feeding which we mentioned earlier. The gist of the method is as follows: definite quantities of corre-

* We shall not return to the discussion of the surface or intracellular localization of oligosaccharidases and oligopeptidases. It is quite clear that if the reader accepts the arguments indicating that these enzymes are localized on the outer surface of the cell membranes, then it will be obvious that their pathology must also be considered as the pathology of membrane digestion.

TABLE 5

Results of Tolerance Tests to Carbohydrates and Lactase Activity in Jejunum in Patients

Data on the patient				Lactose tolerance		Glucose-galactose tolerance		Biopsy of jejunum		
No.	Age (in years)	Sex	Diagnosis	Maximal rise in blood glucose (in mg/100 ml)	Symptoms during test	Maximal rise in blood glucose (in mg/100 ml)	Symptoms during test	Protein mg/g of mucosa	Lactase units/g protein	Alkaline phosphatase units/g protein
1	59	M	Inactive duodenal ulcer	4	Abdominal spasms and diarrhea	57	Absent	64.8	None	325
2	65	M	Benign prostrate hypertrophy	18	The same	24	Abdominal spasms and diarrhea	91.05	5	374
3	56	F	Intolerance to milk (diarrhea)	13 7 (repeatedly)	The same	23 24 (repeatedly)	Absent Absent	95.4 —	2.6 —	248 —
4	40	M	Intolerance to milk (diarrhea), chronic pancreatitis	14	The same	42	Light abdominal spasms	77.9	2.6	423
5	40	F	Intolerance to milk (vomiting), diabetes	21 12 (repeatedly)	Absent Absent	114 —	Absent Absent	97.9 85.1	2.2 3.9 (repeatedly)	550 —
6	40	M	Peptic ulcer, diabetes, intolerance to milk (diarrhea)	41	Abdominal spasms and diarrhea	88	Absent	64.8	5.2	333

Note: In Tables 5 and 6 the units of lactose activity are micromoles of lactose hydrolyzed after 1 min; the units of alkaline phosphatase activity sponding poly- and oligosaccharides which require preliminary cleavage and an equivalent mixture of monosaccharides which does not require digestive treatment are employed; after the administration of the various materials the resulting hyperglycemia permits the solution of the problem regarding the course of digestion of the corresponding polymer. During lactase deficiency the hyperglycemia is compared after the administration of lactose and after the administration of a mixture of galactose and glucose.

We will not cite here the extensive literature regarding lactose intolerance but should like to examine one of the latest and most thorough investigations by Peternel (1965). In this report one can also find numerous literature references.

As can be seen from Tables 5 and 6, the rise in blood-sugar level in individuals not suffering from lactose intolerance following the administration of lactose or glucose-galactose is about the same. The level of lactose in the intestinal mucosa is sufficiently high. In individuals suffering from lactose intolerance the rise in blood-sugar level after administration of the disaccharide is insignificant, while the rise in blood-sugar level after administration of the monosaccharides is very high in many cases. The concentration of lactase in the mucosa approaches zero, and in any case it is several times lower than in normal individuals. It is of interest that in many patients the concentration of another intestinal enzyme, alkaline phosphatase, is not altered.

It is important to note that, judging by the extensive literature, the intolerance to milk may be associated with many diseases of the gastrointestinal tract, and in certain cases there is reason to believe that the lowering of the level of lactase in the intestinal cells is secondary.

In the majority of cases, lactose and other intolerances are not connected with the resorption functions of the intestinal cells, although this is not always the case. As can be seen from Table 5 (patient No. 2), hyperglycemia was absent not only after the administration of disaccharides but also after the administration of monosaccharides. It follows that in several cases there are coupled alterations in membrane digestion and absorption. Apparently, the art of the physician in many cases is reduced to the elimination of the factor which elicits the disease. Indeed, for the preservation of good health it is sufficient that the individuals suffering from lactose intolerance avoid foodstuffs which contain milk sugar. This is comparatively easy for adults but much more difficult for children, keeping in mind the considerably greater distribution of this enzymatic defect in children (Prader and Auricchio, 1965).

TABLE 6

Results of Tolerance Tests Toward Carbohydrates and Lactase Activity in Jejunum of Controlled Patients

Data on patient			Diagnosis	Lactose tolerance test (maximal rise in blood glucose	Glucose-galactose tolerance test (maximal rise in blood glucose, in mg/100 ml)	Biopsy of jejunum		
No.	Age (in years)	Sex				Protein, mg/g of mucosa	Lactose units/g protein	Alkaline phosphatase units/g protein
1	35	M	Functional disturbance of the gastro-intestinal tract	34	73	113	42	186
2	50	M	Hypertonia, peripheral neuritis	24	64	78.6	41.3	671
				69 (repeatedly)	–	–	–	–
3	37	M	Inactive ulcer of duodenum	57	50	58.8	19.5	300
4	73	M	Hernia	98	88	84	121.0	640
5	40	F	Chronic pancreatitis, diabetes	190	No data	125	46.3	282
6	57	M	Doubtful sprue during remission (diabetes)	76	No data	128.5	18.4	160
						101.5	20.0 (repeatedly)	217 (repeatedly)
7	47	M	Ulcerated colitis, resection of stomach (diabetes)	131	102	91.2	44.4	884
8	44	M	Cured ulcer of duodenum (diabetes)	76	94	87.5	15.7	239

TABLE 7

Specificity of Human Intestinal Disaccharidases and Their Quantitative Significance for the Hydrolysis of Various Disaccharides (Dahlqvist, 1962, p. 695)

Enzyme	Substrate	Total activity toward substrate (in %)
Isomaltase (maltase 1a)	Isomaltose	100
	Maltose	50
Invertase (maltase 1b)	Sucrose	100
	Maltose	25
Maltase II	Maltose	15
Maltase III	Maltose	10
Trehalase	Trehalose	100
Lactase	Lactose	100
	Cellobiose	100

Lactose intolerance in newborn children is well known and is recognized late, since nourishment by mother's milk appears to be more adequate and rational. Nevertheless, serious dystrophies can be eliminated comparatively easily by placing the children on a diet free of lactose. The method of administration, described above, as well as aspiration biopsy in these cases reveal the deficiency or complete absence of lactase.

The mechanism of the development of the reaction to milk in individuals suffering from lactose intolerance can be reduced in many cases to the following. Lactose without preliminary hydrolysis penetrates across the intestinal cells only to a very small extent. Thus, in cases of lactase absence practically the entire disaccharide remains in the lumen of the gastrointestinal tract and becomes the "booty" of bacteria. Various metabolites arising in the course of fermentation are absorbed and then elicit the toxic effect.

Bacterial effects in lactase deficiency are developed mainly in the large intestine, although a considerable role is played, apparently, by the flora of the small intestine as well. This is particularly well revealed after surgical alterations by the ileocecal tourniquet (Kalser et al., 1960; Kern et al., 1963).

One patient suffering from lactose intolerance to a small degree had severe reactions after resection of the ileocecal sphincter every time small amounts of milk were administered to her. Kern et al. (1963), who described this observation, are of the opinion that this is connected with the penetration of the cecal flora into the small intestine.

Apparently, the genesis of lactose intolerance in children and adults is different. In children this is related to the alteration in genetic induction of the synthesis of the enzyme, while in adults, at least in certain cases (we know of such a case), the mechanism may be entirely different. During the period of transition to differential nourishment there is a partial repression of the synthesis of β-galactosidase, although the residual activity is sufficient to ensure normal digestion of lactose after the intake of milk. Should the repression be complete, then late lactose intolerance arises. In the case cited above, the milk period was completely normal and intolerance to lactose arose considerably later.

We characterized in detail the picture which develops during consumption of milk or lactose in individuals suffering from a deficiency in intestinal lactase.* Similar events arise during deficiencies in other intestinal disaccharidases (invertase, maltase, isomaltase). Weijers et al. (1960) described three children with chronic diarrhea. On examination of the duodenal contents in these children it was found that the content of amylase was normal. In addition, blood-sugar levels after the administration of various carbohydrates into the gastrointestinal tract were also examined. In one child, whose sugar curve was normal after the administration of a mixture of glucose and fructose or galactose and glucose, a flat curve of blood sugar was observed after the administration of sucrose, maltose, and starch. The authors concluded that in this case there was an alteration of the processes of absorption and a deficiency in maltase and invertase. In the two other children there was a deficiency in one of these enzymes. A certain improvement in the utilization of disaccharides was noted by the administration of the corresponding enzyme preparation together with test breakfast.

Later on many cases of disaccharide deficiency were described (Pander et al., 1963; Lifshitz et al., 1965; see also reviews: Dahlqvist, 1962; Isselbacher and Senior, 1964; Jeffries et al., 1964; Littman and Hammond, 1965).

*A further step toward the understanding of the nature of intolerance to milk was made by Dunphy et al. (1965). During the study of 19 patients suffering from digestive disturbances after intake of milk, the following was observed. In 12 patients the intake of lactose gave a flat curve and the biopsy material contained a sharply decreased quantity of β-galactosidase. At the same time, in the remaining 7 patients intake of lactose was accompanied by a normal rise in blood hexoses; the content of β-galactosidase in the mucosa of the small intestine remained normal. Thus, a thought arises that intolerances to milk include at least two different groups: the first is connected with β-galactosidase deficiency; the causes of the second remain unknown and, possibly, are connected with a specific sensitivity to protein components of the milk, as in celiac disease.

The final stages of the hydrolysis of carbohydrates in the small intestine are performed by a series of disaccharidases. Certain substrates, for instance, trehalose, are cleaved by only one enzyme; others, e.g., maltose, by several. Apparently, if an enzyme is unique, then its absence leads to a sharply defined intolerance. Loss of one of the maltases may be compensated for by the activity of other enzymes (Table 7).

Further purification of these enzymes has shown that in rat small intestine there are at least two invertases and two lactases (see reviews: Isselbacher and Senior, 1964; Prader and Auricchio, 1965).

The deficiencies in disaccharidases, particularly in invertases, may be primary (that is, determined by corresponding genetic defects) and secondary, arising in the course of various diseases such as sprue (Santini et al., 1960), after surgical interferences (Gryboski et al., 1963), during regional enteritis (Prader and Auricchio, 1965), during nonspecific enteritis and virus hepatitis (Sheehy and Anderson, 1965), in cholera and infectious diarrhea of different etiology (Lindenbaum, 1965), and in dystrophy and nonspecific diarrhea in children (Cevini et al., 1962; Bardare et al., 1965). Of interest is the report by Dombrovskaya et al. (1965) concerning carbohydrate intolerances in a considerable number of children suffering from exema.

Often the deficiency in invertase is only temporary. This often occurs in children and could be related to late development of a system which synthesizes invertase in these individuals; this has been demonstrated in rats in our laboratory (Iezuitova et al., 1964c).

It should be noted, however, that in man the isolated deficiency in invertase occurs rarely and in many cases it occurs in combination with alteration in the synthesis of other carbohydrases, most often isomaltase (Auricchio et al., 1965; Prader and Auricchio, 1965).

It may be assumed that diarrhea in children often arises when they are fed substrates which cannot be assimilated because of a retardation in the development of the corresponding enzymatic systems. A practical solution to this problem could be to establish first whether or not the enzymatic systems are sufficiently developed in these individuals before giving them the appropriate food.

We mentioned earlier that membrane digestion is ensured by enzymes of double origin. These are, first of all, intestine-proper enzymes which are synthesized inside the intestinal cells and which are in some manner transferred to the outside and, second, enzymes which are adsorbed from chyme. In this connection a hypothesis was proposed regarding the possibility of an alteration in membrane digestion as a result of an alteration in the adsorption properties of the diseased intestine (Ugolev, 1963). Recently Hooft et al. (1963a, 1963b) made an extremely important observation in this field. While studying child patients after stomach resection these authors observed a malabsorption of starch, although the activity of amylase in the intestinal contents was normal. They were able to observe a correlation between the digestion by amylase and the alteration in the state of the mucosa. At the same time they noted that the mucosa of the small intestine was poorly adsorbing the amylase.

It is possible that in some cases, not only is the reduction in the adsorption of enzymes from chyme of significance but also their incorrect adsorption, as a result of which the enzymes become inactive.

Celiac disease, together with nonspecific intolerances which we considered earlier, may serve as a good example of the pathology of intestinal cells which was induced by nutrients. However, the nature of this induction is different and, possibly, the study of this induction may provide new concepts in the field of physiology of the intestinal cell.

It is certain that at least in many cases celiac disease may be induced by the presence of gluten in the diet and that the exclusion of gluten from the diet may result in significant or complete recovery.

At present it is known that the deamination of glutamine to glutamic acid as well as hydrolysis by papain abolishes the toxic effects of gluten (Krainick and Mohn, 1959). However, after hydrolysis by pepsin and trypsin the toxic properties of gluten are preserved. Apparently, special sensitivity to gluten is essential, since gluten elicits celiac disease only in a small number of people. But when such sensitivity exists, the administration of a solution of gluten into the ileum rapidly induces structural changes typical of celiac disease (Rubin et al., 1962b).

Preliminary incubation of the products of partial hydrolysis of gluten with mucosa of the small intestine of healthy individuals leads, as has been shown by Frazer et al. (1963), to destruction of those fractions which induce the symptoms of nontropical sprue. Undoubtedly, one of the links in the mechanism of this pathological induction is the adsorption of the products of the hydrolysis of gluten on the surface of the intestinal epithelium. Another factor, important to the physiologist, is the connection between atrophy of the villi and the state of the stroma.

Comparison of the situation which is observed during celiac disease and after irradiation damage (see below) leads to the conclusion that the height of the villi is determined by the quantity of epithelium distributed on their surface. In cases of intense desquamation of the epithelium, the stroma is reduced in size with great rapidity. Possibly, this mechanism ensures the uninterrupted epithelial coverage under most varied functional states of the intestine. It could have adaptive significance during a partial loss of the intestinal epithelium since, in all probability, the stroma does not represent an immunological barrier.

We discussed various alterations of intestinal digestion which should be interpreted as a result of this or that defect in the course of hydrolytic processes occurring on the surface of intestinal cells. Additional diseases of membrane digestion become known daily. And one should be prepared for the possibility that this field may become considerably wider than one can visualize today.

Until now we discussed alterations arising in comparatively few diseases, and the initial observations employing modern techniques of investigation of intestinal functions lead one to believe that the circle of these diseases is considerably wider.

It is well known that in many virus diseases such as poliomyelitis, mumps, adenoviral infections, and particularly measles and hepatitis (see review by Sheehy et al., 1964) severe disturbances in digestion with diarrhea and occasionally steatorrhea arise. Such disturbances are known to occur also in viral diseases in animals.

It was recently observed that in many viral diseases considerable structural changes in the small intestine occur. This is true in infectious hepatitis. The fact that enteropaties are very characteristic of viral hepatitis has been demonstrated by direct morphological investigations employing aspiration biopsy (Sheehy et al., 1964). The latter authors expressed the opinion that the alteration of the intestinal epithelium could be an essential and not a secondary component of viral hepatitis. Histological changes, which rapidly developed and disappeared on recovery, were also observed in soldiers with viral enteritis.

Sheehy et al. (1964) attempted to compare the state of mucosa in various viral diseases. They observed fewer significant changes in viral hepatitis than in infectious mononucleosis, rose-rash, chicken pox, and mumps.

It is of interest that pancreatic secretion in severe hepatitis changes very little and even remains normal.

Thus, diarrhea in infectious hepatitis depends, apparently, to a greater extent on the change in membrane rather than cavital digestion. It could be assumed that in measles the digestive functions of the intestinal epithelium may often be altered. The same is true for certain types of adenoviral infections.

There is reason to believe that certain problems connected with the understanding of digestive disturbances in various enteritis and colitis diseases will become clearer when further work in this direction is done using adequate methods for study of the state of the intestinal epithelium.

MEMBRANE DIGESTION DURING IRRADIATION DAMAGE

As soon as it became known that the epithelium of the small intestine of higher animals is one of the most radiosensitive tissues, a considerable number of investigations devoted to the characterization of its state during exposure to irradiation were performed (Conrad, 1956; Perepelkin, 1960; Kurtsin, 1961; Patt and Quastler, 1963; Quastler, 1963). It has been established that with high doses of irradiation (more than 1000 R) the damage to the intestinal epithelium becomes dominant and that they lead to the irradiation death of the intestine. Patt and Quastler (1963) gave a detailed characterization of the damage in a recent review. Apparently, at the base of the alterations in the intestinal epithelium is the extremely high sensitivity of the multiplying elements of crypts toward the penetrating radiation. Cessation of mitosis leads, in the final analysis, to the denuding of the villi, since the defoliated epithelium of the latter is not replaced by the newly formed intestinal cells.

At the present time a detailed characterization is available describing the behavior of cell populations, including the villi during irradiation damage as well as the structure and ultrastructure of the villi and the resorption properties of the intestinal wall. Many such data are collected in Table 8 which is, of course, not complete. Yet, the fact that after irradiation the ultrastructure of the intestinal cells is affected and, in particular, the shortening of microvilli and reduction in their number is observed, furnishes a definite basis for the belief that during irradiation damage membrane digestion will also be affected.

Ugolev et al. (1966) investigated membrane digestion in rats of the Wistar strain (weighing about 170 grams) which were subjected to irradiation employing the apparatus RUM-11 (length of irradiation, 10 minutes; total dose, 1150 R). The activities of invertase, peptidase, and amylase in normal and in experimental rats 4, 24, 48, and 72 hours after irradiation were compared. The rats fasted for 18 hours before the experiment. The activities were determined on inverted pieces of the intestine and on homogenates prepared from similar pieces as well as in the intestinal contents. The defoliation of the epithelium during incubation of the inverted pieces of the intestine in vitro was also determined. Amylolytic activity was determined during perfusion.

Invertase. As can be seen from Fig. 56, the enzymatic activity of the intestinal surface comprised about two thirds of the activity of

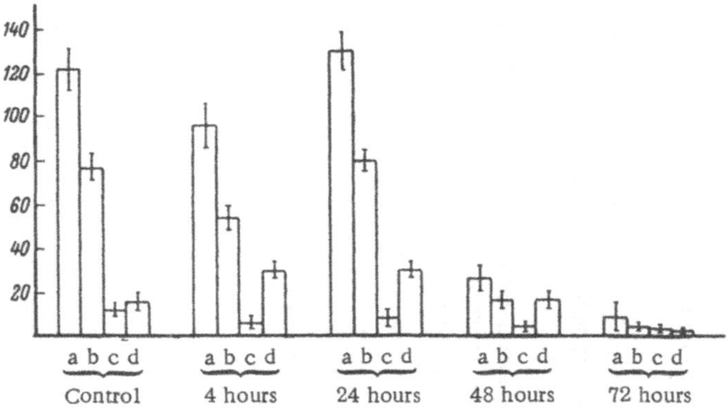

Fig. 56. Effect of irradiation on hydrolysis of sucrose. Invertase activity (in arbitrary units): (a) homogenates; (b) inverted pieces of intestine; (c) solution after incubation of pieces of intestine; (d) contents of the intestine. Ordinate: increase in reducing sugars (in mg % during the time of incubation); numbers below are hours after irradiation.

homogenates in control animals. The hydrolysis of sucrose by the enzymes which pass into solution and accumulate in the intestinal contents is comparatively small (approximately equal to 15-20% of the enzymatic activity of intact pieces of the intestine).

After only 4 hours following irradiation there is a small but definite decrease in the activity of the intestinal surface and a definite increase in the activity of the intestinal contents. The latter circumstance should be considered as a result of increased desquamation of the intestinal epithelium. The decrease in the activity of homogenates during this period is not statistically significant. It is of interest that after 24 hours the activity of the surface and homogenates is restored, although the extent of desquamation remains just as high.

Between 24 and 48 hours there is a sharp decrease in the activity of invertase of intact intestine and homogenates, and after 72 hours the activity of invertase is practically absent.

Peptidase (Fig. 57). The activity of peptidase, as well as that of invertase, is basically determined by the enzymes of the intestinal cells. The activity of homogenates in control animals is approximately equal to the activity of the intestinal surface.

After irradiation the peptidase activity of the inverted slices of the intestine decreases particularly sharply between 24 and 72 hours. After 72 hours, when the activity of the intestinal surface approaches zero, the activity of homogenates is somewhat lowered (statistically insignificantly) compared to normal animals.

The dipeptidase activity in the cavity is sharply increased toward the 72nd hour after irradiation.

Thus, it is characteristic of peptidases that, while the activity of

TABLE 8

Effect of Ionizing Radiation on the Structure and Functions of Intestinal Epithelium

Indicator under study	Dose (R)	Species of animal	Localization of irradiation	Time after irradiation	Character of changes
Absorption of glucose from small intestine	500	Mice	General irradiation	4 hr	Marked decrease of time emptying the stomach, rate of absorption of glucose does not alter. (Source: Dickson, 1955)
				72 hr	Considerable decrease in absorption of glucose. Hexokinase activity in homogenates of the small intestine after irradiation does not change
Time of evacuation of glucose from stomach	700	White rats, line S-D, weighing 120-180 g	Small intestine	From several hours up to 50 days	Increase of time of emptying the stomach after irradiation. Restoration of original level after 8 days. (Source: Moss, 1957)
Absorption of glucose from small intestine	700	The same	Small intestine	72 hr	Decrease in the rate of absorption
	2000	The same	General irradiation with shielded intestinal tract	72 hr	No change in absorption
Comparative absorption of glucose and arabinose in small intestine	700	The same	Small intestine	72 hr	Greater changes in absorption of glucose than arabinose (alteration in the processes of diffusion and phosphorylation). Predominance of alterations in diffusion.

Phosphorylation of glucose	700	The same	Small intestine	24 hr	Inhibition of phosphorylation (amount of organic phosphorus in the mucosa is 1.5 times smaller than in the controls)
Absorption of glucose, fructose, and xylose	1410	Wistar rats, 100-200 g weight	General irradiation	4-20 hr	Decrease in absorption of glucose, fructose, and xylose (in mg per 100 g body weight) from small intestine. In males, significant decrease in absorption of glucose (after 4 hr 14.25, after 20 hr 14.47 against 23.1 in controls). In females, insignificant decrease (15.63 and 14.89 against 14.63 in controls). Large change in the rate of absorption of fructose in females (7.62 and 10.50 against 22.69 in controls) compared to males (12.08 and 9.61 against 10.75 in controls). 4 hr after irradiation in males significant increase (26.28 against 15.94 in controls), after 20 hr certain decrease (11.29) of absorption of xylose. In females 20 hr after irradiation decrease in rate of absorption of xylose, but without sharp rise after 4 hr (9.07 and 15.48 against 14.18 in controls. (Source: Farrar et al., 1956)
Oxidation of glucose by mucosa of small intestine	600	Rats of S-D line, 10-11 weeks old, weight: 185-200 g	General irradiation	3-96 hr	Increase oxidation of glucose 3 hr after irradiation; from 3- to 48-hr level remains steady, on 3-day increase which continues all 4th day and is, probably dependent on bacterial activity and adrenal stimulation. The increase in oxidation of glucose to CO_2 and lactic acid after irradiation is due to increase in activity of citric acid cycle. Irradiation does not affect anaerobic oxidation of glucose (Embden–Myerhof pathway). (Source: Kay and Entenmann, 1959)

TABLE 8 (continued)

Indicator under study	Dose (R)	Species of animal	Localization of irradiation	Time after irradiation	Character of changes
Absorption of water and sodium from cavity of the ileum	2500-3000, 8900	Wistar rats 250-450 g weight	The same	6-48 hr	6 hr after irradiation cessation of absorption from 13- to 48-hr absence of transport of water and sodium across mucosa. After 30-hr cessation of active transport of Na from intestinal cavity to plasma. Assumed direct effect of irradiation on processes of active transport. (Source: Curran et al., 1960)
Absorption of thiamine and histological changes of epithelium of small intestine	525	Rats of CFN line, weight: 155-228 g	The same	72 hr	Significant decrease in intestinal absorption of thiamine up to 17th day. Widespread histological damage in irradiated segments of intestine on 1st day. Cessation mytosis of crypt cells. Resumption of mytosis on 3rd day. Pannett cells are pale and convex. Alteration of absorption of thiamine is due to direct irradiation damage of intestinal epithelium. (Source: Detrick et al., 1963)
Absorption of thiamine and state of epithelium of small intestine (electron microscopy)	525	The same	The same	1 hr–1 day	Absence of changes in transport of thiamine and in thin structure of epithelium of the villi. (Source: Detrick et al., 1961)
				3 days	Lowering of transport of thiamine. Appearance of immature, altered epithelial cells on dehydrated villi. Decrease in size of new epithelial cells

				4-6 days	Microvilli, mitochondria, and cytoplasma of epithelial cells are normal. Partial restoration of intestinal transport of thiamine. Presence of microscopic bleeding
				7-17 days	Gradual restoration of intestinal transport of thiamine. Microvilli are shortened. Microscopic bleeding more intense
Thin structure and function of intestinal epithelium	200, 800, 3000	Mice of Wr line, 2 months old	The same	10 min	Morphological changes in cell membrane. (Source: Quastler and Hampton, 1962)
				1-4 hr	During absorption of corn oil after irradiation by 800 the cells near the top of the villi contained droplets of fat. At 200 and 300 R dose this was not observed. Damage of structure of some mitochondria; nuclei of irregular shape. Space between membranes of irregular shape. Absorption of corn oil almost normal
				2 days	Microvilli are shorter and less regular in shape. In mature cells changes in mitochondria
				4 days	Microvilli of irregular shape, often absent from intact intestinal segments. Mitochondria in mature cells always drastically altered. *Lamina propria* badly infected by bacteria and contains many macrophages. Connective tissue closer to normal than is epithelium. Nuclei altered, enlarged, and irregular in shape; myeloid figures appear in cytoplasm oftener with time after irradiation

TABLE 8 (concluded)

Indicator under study	Dose (R)	Species of animal	Localization of irradiation	Time after irradiation	Character of changes
Cell population of mucose of small intestine	1200	Mice	General irradiation	1-2 hr	In many cells of the crypt picnosis is developed. (Source: Quastler, 1963)
				1 day	Insignificant changes in villi; occasional damage to cell membrane
				2-3 days	Proliferative activity in crypts (synthesis of DNA and mitosis which does not yield daughter cells). Decrease in population of crypt cells and cessation of decrease on 3rd day. Increase in cells and alteration of their structure. Decrease in crypt population after decrease in size of villi. After loss of all cells the submucosal connective tissue becomes swollen; as a result of collapse the villi are transformed into mounds. Due to these changes, a sharp intestinal syndrome is developed resulting in animal death
Structure of intestinal epithelium	2000	Mice	The same	1 hr	Numerous changes in crypt cells: blocking of mitosis. Incorporation of lable into DNA decreases; increase in adsorption in cytoplasm and decrease in nuclei. (Source: Patt and Quastler, 1963)
				2 hr	Picnotic cells and debris of dead cells become convex
				6-12 hr	Crypt cells are damaged more than others. Rate of cell division during this time is reduced to zero

				2-3 days	Crypts are often empty. Rate of loss of cells by the villi is higher than rate of their replacement. Villi are decreased and "disappear." Microvilli, ribosomes, and mitochondria are not numerous and decrease in size; shape is altered. Protein synthesis decreases. Damage to cell structure and function leads to alteration in absorption and to bacterial penetration. Death occurs as a result of considerable "exposure" and bacteremia
Morphologic changes in crypts of the mucosa of the small intestine	1000	Male mice, weighing 18-20 g	The same	Several hours	Lowering of mitotic activity to zero. (Source: Strelin and Galkovskaya, 1964)
				1-2 days	Increase in size of crypt cells by 30% with decrease in their number by 40%
				3 days	Repair
				10-15 days	Death of some animals due, apparently, to alteration in production of blood (majority survived 1 month)
	2000	The same	The same	1-2 days	Increase in size of crypt by 50% and reduction in their number by 60%
				2-3 days	Comparatively low mitotic activity
				4 days	Repair is not quite complete
				4-5 days	Death of animals

the intact pieces of the intestine falls, high activity of homogenates and increased hydrolysis by the enzymes which pass into solution after incubation of the inverted pieces of the intestine prevail.

Amylase (Fig. 58). First of all, a considerable variability of amylolytic activity in the passing perfusate and in the intestinal contents should be noted. This seriously limits the possibility of describing the changes which takes place after irradiation. Apparently, the following facts are worth noting. The amylolytic activity of the contents of the small intestine is sharply increased 24 hours after irradiation. Nevertheless, hydrolysis of starch on the surface has a tendency to decrease. After 72 hours, when the activity of the contents of the small intestine is somewhat higher than in control animals, the hydrolysis of starch during perfusion is close to zero.

Thus, the data obtained permit the assumption that the defects of membrane digestion are of definite significance in the pathogenesis of alterations which arise after severe irradiations.

Indeed, the considerable experience of clinicians who studied various intolerances suggests that the sharp decrease or absence of activities of definite enzymes results in the fact that the corresponding polymers and oligomers become an easy "prey" of the bacteria present in the intestine leading to all the consequences which were characterized during the description of intolerances.

Under these conditions the cases of nutrient alterations resemble oligomer intolerances because cavital digestion is altered to a considerably lesser degree (see reviews: Perepelkin, 1960; Kurtsin, 1961). Therefore, the initial stages of hydrolysis take place without significant deviations. The comparatively greater stability of the processes of cavital digestion is understandable if one keeps in mind that the truly secretory processes are performed by cells which are much less sensitive to irradiation than the intestinal cells.

Let us now examine in detail the alterations which we described above.

First of all, it is of interest to note the initial inhibition of the enzymatic activity after 4 hours which is compensated for after 24 hours, and that the processes of active transport are progressively inhibited with time. How should we interpret these differences in the reaction of digestive and resorption processes performed by the intestinal cells? Apparently, the activity of the already synthesized enzymes distributed on the surface of the cell membranes, in contrast to processes of active transport, is less dependent on the energetics of the cells. Certain data relating to the alterations in phosphorylation are listed in Table 9.

Of interest is the nonidentity of the reactions of dipeptidase and

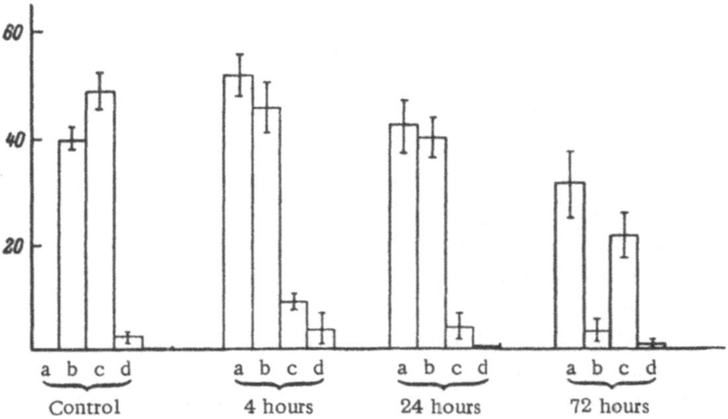

Fig. 57. Effect of irradiation on hydrolysis of glycyl-L-leucine. Ordinate: dipeptidase activity (in arbitrary units). The rest of the notation is the same as in Fig. 56.

invertase which are synthesized by the same cell elements. Indeed, the inhibition (to complete) of invertase activity in homogenates as well as in the intact intestine pieces indicates an alteration of the synthesis of this enzyme and, perhaps, its transfer (translocation) to the cell surface.

Apparently, in the case of dipeptidases, basically the second

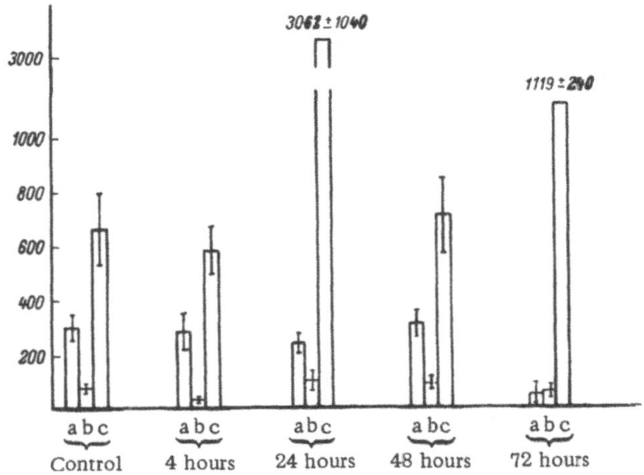

Fig. 58. Effect of irradiation on hydrolysis of starch. Ordinate: amylolytic activity (in % of hydrolyzed starch in 1 minute). (a) Rate of hydrolysis during perfusion; (b) amylolytic activity of the passing perfusate; (c) amylolytic activity of the contents of the small intestine before perfusion. The rest of the notation is the same as in Fig. 56.

process is affected, namely, the translocation of the enzyme, since the total amount of the enzymes present in the intestinal cells does not significantly alter. It seems to the author that the data regarding the sharp inhibition of dipeptidase activity is of interest in still another respect.

As was mentioned earlier, after 48 hours the level of dipeptidase activity of the intestinal cells remains quite high, although the intact intestine practically does not hydrolyze glycyl-leucine. Thus we must again come to the conclusion that the presence of enzymes in the intestinal cells and the digestive functions of the intestinal surfaces are far from being identical phenomena. In our case the enzymes present in the cells were inactive.

During the investigation of the cavital and membrane hydrolysis of starch we were confronted with phenomena which resembled those described by Hooft et al. (1963b) in children with resectioned stomach. In an entire series of cases, in spite of the high amylolytic activity of the intestinal contents (even higher than in the normal state) the capacity of the intestinal surface to hydrolyze starch was nearly zero. Thus, during irradiation sickness one also observes the weakening of the processes of adsorption by the intestinal cells. However, in certain experiments we were confronted with phenomena which can be interpreted not as a weakening of adsorption but as an incorrect adsorption of the enzyme on the surface of the cells.

As can be seen from Fig. 59, during perfusion the hydrolysis of starch in vivo was extremely low; this indicates a weakening of membrane digestion, and yet in the passing perfusate the level of amylase was very high. On comparison of the two facts the conclusion is inevitable that the amylase was adsorbed but was inactive either because the conditions in the region of the surface were inadequate (for

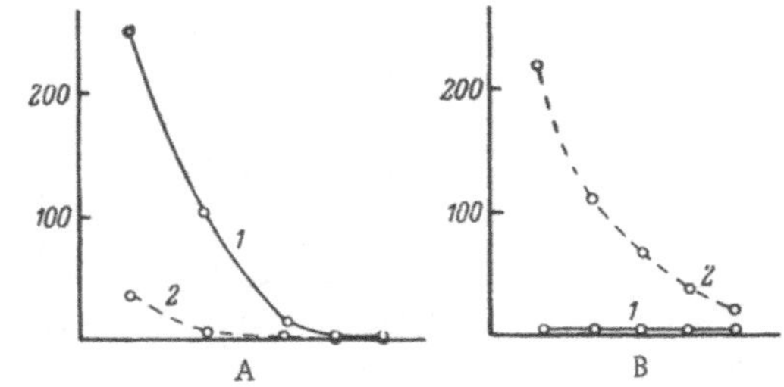

Fig. 59. Correlation of the hydrolysis of starch in the intestine (1) and in the passing perfusate in vitro (2) in a healthy animal (A) and 72 hours after irradiation (B). In the latter case one observes adsorption of amylase with loss of activity. Ordinate: amylolytic activity (in arbitrary units).

instance, the charge may have been altered) or because the molecule of the enzyme was adsorbed in such a way that the active center became screened off. This question requires a special solution, but in general a conclusion should be drawn that in certain cases the enzyme becomes inactive and that after its desorption in solution its activity is restored. Apparently, we are confronted here with one more possibility in the pathology of membrane digestion.

Everything that was said above relates to the initial stages of severe irradiation damage. So far there is no information regarding the state of membrane digestion during the period of restoration of the functions of the intestinal tract. It is clear that such information would be of considerable interest.

If we set aside the clinical problems involved in irradiation damage and consider irradiation damage only as a model, then it would appear that this model is very convenient for the characterization of various alterations in membrane digestion.

Thus, at the present time we know of various forms of pathology of membrane digestion which could arise as a result of alteration in the ultrastructure of the intestinal surface, the composition of the enzymatic layer of the intestinal surface, and in the changes of its adsorption properties. It is not excluded that in the not so distant future a series of other forms of pathology will become known which at present are difficult to predict. Theoretically, the alterations in membrane digestion as a result of weakening of those forms of intestinal peristalsis which ensures a steady transfer of substances from the chyme into the brush border appear to be quite possible. In addition, it is possible that during the alteration in the state of the membrane of the microvilli, the access of foodstuffs to the pores of the brush border may become difficult.

During the analysis of various alterations of the digestive functions of the small intestine it was noted that the presence of enzymes on the surface of the membrane is ensured by: synthesis in the ribosomes and by a mechanism which ensures the transfer of enzymes on the membranes of the microvilli (most likely as components of lipoprotein structures) (Timofeeva et al., 1965; Ugolev et al., 1966).

In many cases the digestive functions become altered because the synthesized enzymes are not translocated to the surface.

Sterk and Kretchmer (1964) compared the activity of β-galactosidase of rabbit small intestine immediately after birth with that following birth. They found that the fraction of enzymes acting on the surface decreases with age. It is thus possible to assume that the repressed capacity to cleave lactose is not determined by repression of synthesis but by inhibition of the translocation of β-galactosidase onto the surface of the brush border. It is possible that this is connected with the fact that other enzymes begin to compete for the possession of the "enzymatic points" of the cell membrane, and that the synthesis of these enzymes begins in connection with the transition from the milk nourishment to a more definitive one.

What are the consequences of alterations in membrane digestion to the organism? Insofar as membrane digestion is a supplier of the products of absorption one could expect that its pathology will be accompanied by an alimentary deficiency. We have pointed out that the enhancement of cavital digestion during an insufficiency in membrane digestion cannot always be considered as a favorable factor (Ugolev, 1963). Indeed, if hydrolysis in the cavity proceeds to the formation of end products such as glucose, amino acids, etc., then extremely favorable conditions arise for the nourishment of bacteria populating the small intestine. Wild multiplication of microflora under these conditions may aggravate the state of the patient.

If such a schematization is permissible then one may say that during a sharply reflected deficiency in membrane digestion the cavital digestion is transformed into a mechanism which ensures the nourishment not of the macroorganism but of bacteria which populate the gastrointestinal tract. Is it due to this that starvation favorably affects individuals suffering from enteritis? Clearly here we speak only about certain characteristic aspects of the state of the patient in whom membrane digestion has been altered.

COMPENSATORY REACTIONS

In the characterization of the clinical aspects of membrane digestion we should examine not only those cases in which an alteration of this mechanism exists but also those in which membrane digestion acts in the role of a compensatory process. It is logical to ask: can the deficiency in cavital digestion be compensated by membrane digestion within definite limits? We have cited a series of facts supporting this possibility (Ugolev, 1963). Insofar as the transition from hydrolysis in chyme to hydrolysis in the pores of the brush border of sufficiently small molecules is a probable mechanism, it could be expected that the decrease in the tempo of cavital digestion will result in the arrival of larger molecules at the brush border. This, as we have already mentioned in one of the preceding sections, will produce a very significant compensatory effect.

Since the direct facts regarding the compensatory function of membrane digestion are still few we shall cite a series of indirect data.

Pavlov and many later investigators (Goldberg, 1913; Douglas et al., 1953; Johnson, 1953; Jackson, 1964; and many others) showed that the animals in which the entry of the pancreatic juice into the small intestine was excluded in one way or another (pancreatectomy, ligation of pancreatic ducts, permanent pancreatic fistula) do not suffer serious digestive disturbances. Particular attention should be directed to the absence of alterations in the digestion of carbohydrates, since the pancreas appears to be the only organ which ensures the synthesis of amylase which is involved in the transformation of polysaccharides. The absence of the pancreatic juice is partially compensated for by a more prolonged stay of the food in the stomach, although of considerably greater significance in the processes of compensation, apparently, is the intestine with its capacity to adsorb and accumulate on its surface various enzymes present in the chyme.

Various authors have shown that the adsorptive effects of the intestine in regard to pancreatic amylase are particularly evident when the concentration of the enzyme in the medium surrounding the intestinal cells is small (Ugolev, 1960b, 1961a; Lee, 1962).

During deficiency in cavital digestion caused for instance, by the resection of the stomach, there is a structural reconstruction of the intestinal mucosa which, possibly, ensures the intensification of membrane digestion. The effect of cavital digestion, weakened for any reason, may be increased by increasing the length of stay of food in the small intestine. This phenomenon has been described in detail by Filipovitch (1962) and Filipovitch et al. (1960, 1961) in dogs with resectioned stomachs. The reconstruction of the motor activity of the intestine followed by stomach resection, exhibiting compensatory character has been noted by Gordon et al. (1959) and by Kulyukin (1964) and by others. Apparently, very useful in this connection will be investigations employing electron microscopy and approaches which permit studies of the enzymes on the surface.

In what manner can the compensation of cavital digestion be accomplished by means of membrane digestion?

Earlier we mentioned certain theoretically possible ways. These are the changes in diameter of the pores of the brush border, an increase in the adsorption properties of the intestinal epithelium, and, finally, changes in the proximal–distal gradient at the expense of the increase in digestive functions of the distal segments of the small intestine which, under normal conditions, contain smaller amounts of the enzymes (see reviews: Spencer, 1960, 1964a; Ugolev, 1963).

Of great interest is the possibility of compensation of a deficient function of certain segments of the small intestine by the others. Thus, Bochkov (1958) observed that after partial (1/3-1/2) resection of the intestinal tract in dogs, there is a widening of the diameter of

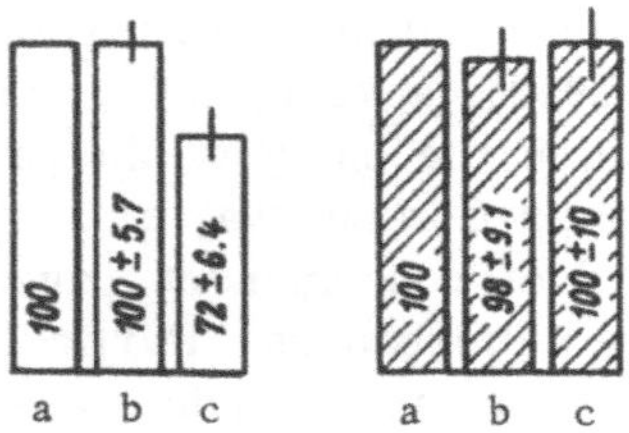

Fig. 60. Changes in the proximal–distal gradient of maltase activity in the rat small intestine with 10-week-long lead poisoning. Light columns: control; shaded columns: experimental. Activity of the middle (b) and distal (c) segments of the intestine is expressed in percent of activity of the proximal (a) segment, taken as 100%.

the remaining portion, thickening of the wall of the intestine by 20-30% at the expense of hyperplasia of the cell elements of the mucosa. In scrapings of the mucosa in these animals there was a considerable increase in enterokinase beginning from the 2nd or 3rd week following the resection.

Changes in the proximal–distal gradient under normal and certain pathological conditions were observed in our laboratory. Nadirova (1965) investigated the enzymatic function of the small intestine in rats with chronic experimental lead poisoning and observed that during alterations of digestion in proximal segments of the small intestine there is a compensatory rise in the activity of certain enzymes (maltase and invertase) in distal segments (Fig. 60). True, in this case we are not speaking about a compensation of deficiency in cavital digestion but about a replacement of membrane digestion in certain segments by other segments. Nevertheless, these data are of interest as a general case. Since this particular investigation contains important details we shall dwell on it a bit more thoroughly.

The investigation of the digestive properties of the small intestine was performed by the following means: (1) comparison of the increase in blood hexoses after administration of glucose and maltose into the upper segment of the duodenum (the technique and certain data obtained under normal conditions were described on pages 105-115); and (2) direct determination of the enzymatic activity in vitro employing inverted sections attached to glass rods obtained from various segments of the small intestine.

In order to characterize the intensity of membrane hydrolysis of disaccharides, it appeared logical to introduce the concept of the maltose–glucose coefficient which shows the ratio of the rates of absorption of hexoses after administration of equivalent quantities of maltose and glucose.

We have already mentioned that under normal conditions the rate of absorption of hexoses into the blood 15 and 30 minutes after the administration of glucose and certain disaccharides is the same, that is, that membrane digestion is not a factor which limits the assimilation of disaccharides.

During lead poisoning (length of poisoning, 2 weeks) the picture essentially alters. The maltose–glucose coefficient 15 and 30 minutes after the administration of the corresponding carbohydrates in experimental animals is half as large as in the controls:

	After 15 minutes	After 30 minutes
Control animals	1.32	0.82
Animals with lead poisoning	0.62	0.43

It must, therefore, be concluded that the cleavage of disaccharides takes place at a diminished rate; that is, at certain stages of the development of lead poisoning there occurs a weakening of membrane digestion.

It would be simpler to assume that such a weakening of membrane digestion is determined by the diminution in the amount of enzymes. However, in experiments in vitro at earlier stages this has not been observed. If we now depart from the direct data and attempt by means of risky hypotheses to understand the reason for the diminution in the cleavage of disaccharides in the proximal segments of the small intestine of the poisoned rats, then it would appear possible to assume that here we are dealing with an alteration in membrane digestion as a result of damage to the motor function. Indeed, a series of investigators have shown that in lead poisoning the intramural nerve apparatus of the small intestine is damaged (Mambeeva, 1960; and others) and, also, there has been a direct demonstration of the alteration of the motor function of the small intestine (Levrat et al., 1954; Kaliteevskaya, 1957; Dinischiotu et al., 1960; Brenner, 1962; Mambeeva, 1963; and others).

Regardless of whether this hypothesis is correct or not, it is clear that a weakening of the tempos of digestion of disaccharides must result in the entry of the latter into more distal segments of the small intestine creating thereby conditions for induction by the food substrates of the corresponding enzymes.

Indeed, as can be seen from Fig. 60, there is a significant and considerable increase in the enzymatic activity in the distal segments of the small intestine.

It should be noted that such a phenomenon also occurs under normal conditions and it will be characterized in the description of ontogenesis. In very young rats it is the distal segments of the small intestine which possess the highest carbohydrase activity. During transition to the definitive type of nourishment the maximum activity is shifted to the proximal segments. At that time we expressed the hypothesis that such a redistribution of enzymatic activity is connected with the blocking of the surface of the proximal segments by the surface-active lipids. Therefore, at the early stages of ontogenesis an increase in the enzymatic function of the distal segments also could be connected with the fact that the substrates which induce the synthesis of the corresponding enzymes penetrate far up along the intestinal tube.

It is possible that a similar mechanism is at the base of the reconstruction of the proximal–distal gradient in man suffering from celiac and other diseases. Thus, Schedl and Clifton (1963) observed a more intensive absorption of glucose in distal segments of the small intestine in patients with nontropical sprue, which corresponded with the absorption of glucose in proximal segments of healthy individuals. It would appear that in this disease the distal segment of the small intestine compensates for the deficiency in the function of the proximal segment which, as is well known (MacDonald et al., 1964a; and others), is much more affected in nontropical sprue. Since there is a close correlation between the proximal–distal gradient of absorption and membrane digestion, it may be assumed that in nontropical sprue there exists a compensatory increase in the digestive activity of the distal segments of the small intestine.

The possibility of a direct induction of the synthesis of enzymes by intestinal cells will be considered later.

The clinical aspects of membrane digestion are a much greater problem than the problem of diseases connected with its alteration. Here we must include such important questions as, for instance, the effect of various surgical interferences with the stomach, pancreas, and other organs of the gastrointestinal tract on the processes of membrane digestion. The question arises whether one can assume a priori that conpensatory reactions will occur which will compensate for the dificiency in cavital digestion or, on the contrary, should one expect that even this important process will be altered.

The available literature data indicate that after partial resection of the stomach one may observe the first, that is, a compensation, and the second, namely, alterations in digestive functions of the small intestine, its structure and ultrastructure which arise after a variable period following the operation (see reviews: Filipovitch, 1962; Jeffries et al., 1964; Dumesh, 1965). In the final analysis, we are dealing here with the fact that the evaluation of the general state of the digestive system and the selection of the most rational tactics must include digestive processes which take place on the surface of the intestine and which play an important role in the general balance.

Although the quantitative information regarding alterations in the digestive functions of the intestinal cells is at present not very extensive, it may be stated with assurance that the most interesting still remains unknown.

A question arises: why is it that during irradiation damage there is an alteration in the synthesis of invertase, whereas peptidase is synthesized in amounts close to normal but is not transferred to the surface of the intestinal cells? It is comparatively easy to understand disaccharidase deficiencies in celiac diseases, but it is not clear why the synthesis of invertase and lactase is altered in lambliasis (see review by Prader and Auricchio, 1965). It is of interest that a deficiency in the enzymes accomplishing the final stages of hydrolysis of carbohydrates which arises in diseases of the small intestine disappears in the course of treatment, but it is not clear why lactase activity is restored much more slowly than the ability to cleave other disaccharides.

We know very little about what takes place within the limits of one intestinal cell, such knowledge would be insufficient for understanding alteration of membrane digestion and its compensatory reactions.

In various chapters of this book we discussed the behavior of all the elements of the villus as a single unit. There is ample basis to assume that the course of the final stages of hydrolysis and active transport of substances in several instances are determined not so much by the alteration (or not only by alteration) of functions of individual cells as of the villus as a whole. From this standpoint the alteration of the structure of the villi and of the kinetics of renewal of cellular material (as well as its aging) in celiac diseases is of great interest (see reviews by Padykula, 1962; Jeffries et al., 1964; and others). It is possible that certain alterations of membrane digestion are induced at the free surface of the intestinal cell, for instance, glutenogenic forms of sprue, while the others are induced from the side of the stroma. Finally, the reactions of the small intestine as a whole are even more complicated. We cannot answer any of the questions which we raised. Furthermore, there is no satisfactory theory of pathogenesis of these diseases because as far as we know, there is no satisfactory concept which characterizes the regulation of the processes which take place within a single intestinal cell or in a population of intestinal cells under normal conditions.

Chapter VI

Evolutionary Aspects of Membrane Digestion

The views developed in this book can, in general, be reduced to the following: there are, at least in higher animals, three fundamental stages of operation of the digestive apparatus—cavital digestion which performs the initial transformations; membrane digestion which completes the cleavage of foodstuffs; and, finally, the processes of absorption.

Until now we have not analyzed the peculiarities of the digestive processes in various species of animals, in animals which are at various stages of the evolutionary ladder during different periods of ontogenetic development.

In this chapter we examine the evolutionary aspects of membrane digestion insofar as it is possible on the basis of available information.

We mentioned earlier that membrane digestion is a widely distributed mechanism and, perhaps, a universal one among the heterotrophic organisms beginning with the most primitive and ending with man. The discussion of this view at the present time, compared with previous summaries (Ugolev, 1963, 1965a), is based on a larger number of reports which consider additional groups of animals. It is even more important that new information became available regarding the species peculiarities of membrane digestion which, apparently, are just as diverse as in the case of distal extracellular digestion.

It often becomes possible to connect membrane digestion with the peculiarities of nutrition of a given species of animals. In other cases, at least at the present time, this is impossible. Nevertheless, even today the question of adaptive evolution of membrane digestion is becoming of central importance in ontogenetic as well as in phylogenetic aspects.

TABLE 9

Functional Indicators of the Small Intestine during Prenatal Period (Spencer, 1964b, p. 108)

Functional indicators	Object of study	Notes
Acid phosphatase	Chick	Present in tissues at early stages of embryonic development
	Swine	Higher at birth than in prenatal period
Active transport: 6-deoxy-D-glucose, galactose, and L-histidine	Rabbit	Present in the gut of offspring and its detected under anaerobic conditions
Glucose and galactose	Chick	Absent in embryonic tissue of the gut (develops postnatally)
Alkaline phosphatase	Man	Present before birth
	Chick, mouse, guinea pig, trout, other species	Present in the gut of offspring, activity changes with time. Enzymatic activity changes under the influence of adrenosteroids and thyroid hormones
Amylase	Man	Present before birth
Cholineesterase	Chick	Present in cultures of intestine of embryonal chicks; its synthesis inhibited by puromycin
Enzymes: asparogenase, maltase, invertase, "erepsin," lecithinase, glycyl-tryptophane peptidase, enterokinase, glycerophosphatase. Presence of secretin	Man	Early literature (up to 1930, collected by Needham (Chemical Embryology, III, New York, 1931)
Enterochromatin cells	Man	Cranial—caudal gradient present in differentiation subsequent distribution of enterochromatin cells
Enterokinase	Man	Present from 30 weeks on
Enzymes: reaction of the gut of offspring to epinephrine	Cow, sheep, dog, swine, horse, cat	Early literature (collected by Needham, see above)
Morphology of epithelium	Rat	Pinealectomy during prenatal period at 15-17 days elicits changes in ileum from the first week
Esterase, nonspecific β-galactosidase	Chick	Increase after 9th day of incubation
	Rat, rabbit	Present, reaches highest activity at birth

TABLE 9 (concluded)

Functional indicators	Object of study	Notes
Invertase	Man	Present early
	Rat, mouse	Absent (develops postnatally). Invertase is detected when hydrocortisone is added to intestine culture of 5- to 6-day-old rats
Lactase	Man	Present before birth
	Swine	Present before birth in low concentration
Lipase	Man	Present before birth
	Swine	Lower concentration in early embryonic period than in adults
Maltase	Man	Present from 22nd week on
Peptidase	Man	Present before birth
"Saccharase"	Man	Present from 10th week on
Secretin	Man	Probably present in low concentration at birth
Succinedehydrogenase	Chick	The enzyme present in duodenum from 15th day. Not increased significantly by cortisone administration. Thiourea inhibits development of enzyme
Trimetaphosphatase	Chick	Absent (data available for other species)
Incorporation of DL-methionine-H-3 and DL-ethionine-H-3	Rat	Present in the intestine of offspring after injection of these amino acids
Vitamin B_{12}	Man	Slightly higher in embryos of early embryonal period than later on in newborn children
Ratio of the weight of small intestine to that of entire tract	Cat	Increases in embryos between early and late embryonal period
Functional indicators	Object of study	Notes

TABLE 10

Functional Indicators of the Small Intestine during Prenatal Period (Spencer, 1964b, p. 108)

Functional indicators	Object of study	Notes
Absorption of acetic acid from rumen	Cow	Greatest absorption in calves receiving supplemental food; enzymes studies were performed on developing rumens; the development of rumens decreased with decrease in weight
Active transport of L-histidine	Rabbit	Transport capacity under anaerobic conditions decreased during the 1st week of postnatal life, while aerobic transport was increased
Glucose and galactose	Chick	Present within 2 days after birth
Absorption of α-amino-isobutyrate	Rat	Concentrating capacity of duodenum drops with age, but the weight and growth of the animal is more important than age, as was shown in underfed animals
Acid phosphatase, alkaline phosphatase	Swine	Higher during birth than during prenatal period
Alkaline phosphatase, absorption of antibodies	Rat	Deoxycorticosterone increases the activity of the enzyme and decreases the absorption of antibodies in newborn rats
Amylase	Swine	Approximately constant between the 1st and 5th week of postnatal development
Bacterial content	Swine	Bacterial invasion evident after few hours following birth; permanent population at 3-4 weeks of age
Absorption of calcium	Rat, swine	Higher in young animals compared to old ones
Absorption of cholesterol	Chick	Absorption of cholesterol "reaches maturity" in cocks weighing 1900 g
Development and proteolytic activity of middle segment of the intestine	Larva *Tribolium castaneum*	Soybean trypsin inhibitor inhibits proteolysis and development of the larva
Digestion	Swine	Colostrum of swine contains trypsin inhibitor; probably, it plays a role in preventing cleavage of immune proteins. Colostrum of man and cow also contains immune proteins
	Man	For deductions regarding digestion in newborn see Platt, 1961

TABLE 10 (continued)

Functional indicators	Object of study	Notes
Enzymes	Man	Enterokinase, alkaline phosphatase, "saccharase," peptidase, lipase, amylase alter in amounts and are present before and after birth
Nonspecific esterase	Chick	Reaches level of adults after birth; large increase in enzymatic activity is observed between the 1st and 10th days
β-galactosidase	Rat, rabbit	Highest activity at birth, decreases thereafter
Absorption of glucose and alkaline phosphatase	Rat	During postnatal period their development has no definite distribution. Absorption of glucose in rats on 10th day after birth is stimulated by ACTH and cortisone. On the 9th day the concentrating capacity of the middle segment of the intestine is higher than in other segments
Absorption of immune proteins	Swine	Absorption from colostrum decreases logarithmically after birth and ceases at 24 hr of age
pH of the gut	Swine	In 18-month-old animals the pH is about 0.4 unit higher than in 5-month-old animals
Invertase	Rat, mouse	Present from 2nd postnatal week. Could be stimulated by hydrocortisone right after birth
Lactase	Swine, cow	Present in very young animals; enzymatic activity may then fall, but in swine intestine it increases
Lactase, invertase, maltase	Man	Could be inborn absence in the small intestine
Lipase (tributyrinase)	Rat	Increases between 10th and 18th day after birth (time of weaning)
Maltase	Cow, swine	Low at birth; may increase
Peristalsis	Rat	Increase during 3-7th day of postnatal period, then again between 18th and 21st day
Absorption of polyvinylpyrrolidone (PVP)	Swine, sheep	PVP is absorbed during the first 24-36 hr of life. During starvation the absorption of PVP continues up to 86th hr. Cow's milk induces cessation of absorption
Proteinase	Swine	Increases after 2nd postnatal week

TABLE 10 (concluded)

Functional indicators	Object of study	Notes
Formation of salicylic acid glucuronate	Guinea pig	Increases approximately 4-fold between birth and the 3rd day of postnatal life. Continues to rise thereafter up to adult level
Succinedehydrogenase	Chick	Increases during the first 5 days after hatching, then falls
Saccharase	Swine	None at birth, then increases
	Rat	By 23rd day is 2/3 of level of activity in adults
	Ox	Not detected
Absorption of vitamin B_{12}	Rat	Absorption is independent of intrinsic factor and falls after birth, then reaches adult levels by the 14th day of postnatal life
Vitamin C	Man	Lower in newborn than in embryos of early embryonic period

MEMBRANE DIGESTION IN ONTOGENESIS

Beginning in the second half of the last century, the development of the functions of the digestive system (a problem equally important to theoretical biology and to applied fields) has been studied very intensively (see reviews by Gartier, 1900; Windle, 1940; Arshavskii, 1957; Smith, 1959; Platt, 1961; Wilson, 1962). Extensive information of a descriptive nature has been accumulated (Tables 9 and 10).

Yet, it is much more difficult to understand the mechanism of the work performed by the digestive apparatus during the period of postnatal development of mammals and man when the primary foodstuff is milk. It should be stated that this problem consists almost entirely of questions to which there are no answers. Indeed, it has been shown that in newborn mammals the secretion and synthesis of stomach proteases are very weak (Windle, 1940; Hill, 1956; Arshavskii, 1957; Kravitskaya, 1959; Smith, 1959; Mosinger et al., 1959; Platt, 1961; and many others), the activity of pancreatic lipase is low (Kruchkova, 1939; Rokos et al., 1962a, 1962b; Kuznetsova, 1963), and the activity of amylase approaches zero (Klumpp and Neale, 1930; Kruchkova, 1939; Smith, 1959; Stepanova, 1960; Platt, 1961; Kuznetsova, 1963; Marauska et al., 1963; Salenietse et al., 1963).

Figures 61 and 62 show data which characterize the secretion of amylase of the pancreas in rabbits and pups during various periods

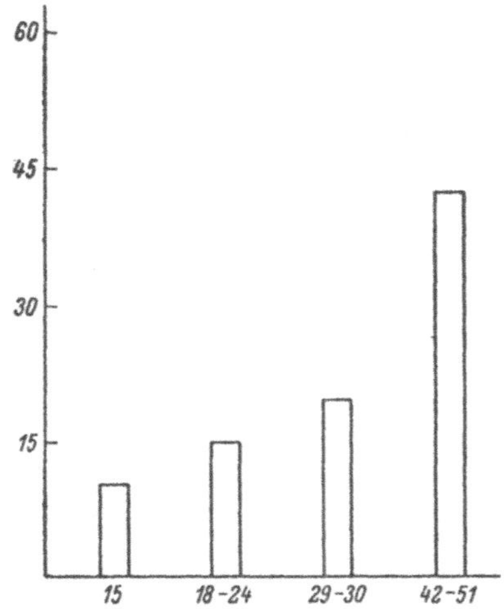

Fig. 61. Amylolytic activity of the pancreatic juice in pups of various ages (milk fed) (according to Kuznetsova, 1963). Abscissa: age of animals (in days); ordinate: concentration of amylase per ml of juice (in arbitrary units).

of ontogenesis (Kuznetsova, 1963; Ugolev and Salenietse, 1964). As can be seen from the figures, the development of the cavital digestion of starch is adapted to the moment of transition from milk nourishment to the definitive one.

The deduction made by numerous investigators that cavital digestion during milk nourishment is weakly developed appeared quite justifiable. However, a question arises: in what way are the foodstuffs assimilated during this period? Various assumptions were expressed. Of these the following deserves the greatest attention.

Milk contains a certain amount of its own enzymes which aid the hydrolysis of food. It does not require any enzymatic pretreatment. However, these enzymes of milk do not, apparently, play a significant role, since their inactivation does not result in noticeable changes in the rate of utilization of foodstuffs by the newborn organisms. As regards the possibility of penetration of the intact components of milk into the inner medium, this process may be of some significance during the first few days of postnatal development when, according to the data of Clark (1961a, 1961b) and others, the intestinal cells possess a definite pinocytotic activity (see reviews by Platt, 1961; Wilson, 1962). It is not clear, however, to what extent the pinocytosis, which aids the passage of uncleaved substances across the intestinal barrier and intracellular digestion, can ensure the nutrient requirements of the organism. Besides, soon after birth the intestine begins to function as the immunological barrier, and pinocytosis is reduced to a minimum, as has been shown by means of electron microscopy (Clark, 1961a, 1961b).

Thus, in the majority of mammals there is a more or less prolonged period of milk nourishment, when pinocytotic activity is very low and

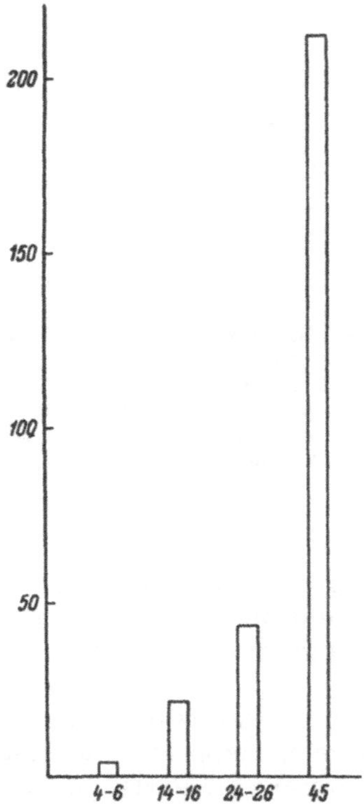

Fig. 62. Rate of secretion of amylase per minute during various periods of ontogenesis in rabbits. Abscissa: age (in days); ordinate: amylolytic activity (in arbitrary units).

enzymatic activity of the contents of the stomach and intestine is not yet developed to any significant extent.

From purely theoretical considerations we expressed a hypothesis (Ugolev, 1963) that during this period, which is characterized by nourishment by highly dispersed food which does not require significant preliminary enzymatic treatment in the digestive tract, the basis hydrolysis is performed by membrane digestion. This hypothesis has been intensively verified in the past few years in our laboratory and in others. The available data, apparently, are adequate to give an affirmative answer.

We shall now turn to a more detailed examination of the problems regarding the role of membrane digestion in the early postembryonic period.

In the first investigations in this direction (Marauska et al., 1963; Ugolev and Salenietse, 1964) the hydrolysis of soluble starch was investigated in rabbits during perfusion of the small intestine (including the duodenum) keeping the entry of pancreatic juice into the perfused segment intact. The animals were studied on the 4th to 6th, 14th to

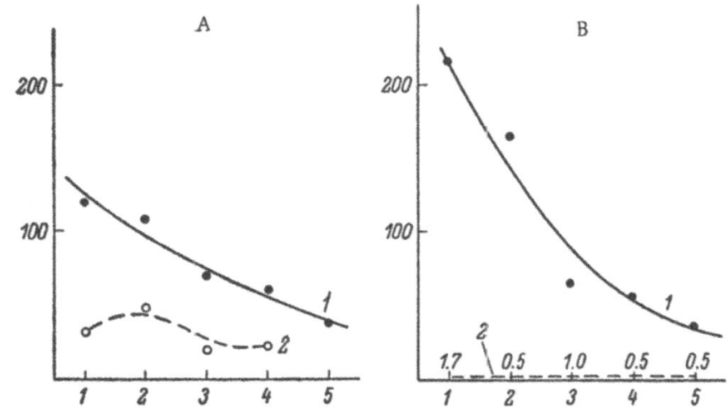

Fig. 63. Hydrolysis of soluble starch in rabbits on the 45th (A) and the 4th (B) day of postnatal development. Abscissa: trial number; ordinate: intensity of hydrolysis (in arbitrary units). (1) Hydrolysis in vivo; (2) in vitro. In B the curve was drawn arbitrarily, since the intensity of hydrolysis (numbers on curve 2) was close to zero, and it was difficult to reproduce the curve on the scale used in the figure. The trials were performed every 3 minutes.

16th, 24th to 26th, and 45th days of postnatal life. These data were compared with those obtained on adult animals using the same methods.

As can be seen from Fig. 63, the hydrolysis of starch in the intestine of a 45-day-old rabbit is similar to that described for the adult animal. The hydrolysis in vivo (that is, during the time the substrate remained in the intestine) was 2-5 times higher than the hydrolysis in vitro (that is, under the influence of enzymes which pass into the perfusate). This indicates that only a certain portion of the substrate was cleaved by the enzymes acting in the cavity, and that a considerably greater amount was hydrolyzed by the enzymes fixed on the surface of the intestine.

The level of hydrolysis in vivo of the 4-day-old rabbit is not lower than the level of hydrolysis in animals which were on an independent feeding regimen. Of interest is the fact that the passing perfusate almost does not contain amylase, while in a 45-day-old animal this enzyme is present in considerable amounts.

Figure 64 illustrates data describing the results of the experiments performed by Salenietse and the author. As can be seen, in 4-day-old rabbits as well there is a weakened hydrolysis of starch in the course of perfusion as a result of desorption of amylase from the surface of the intestine. Certain peculiarities of this process in newborn animals compared to that in adults, observed in the experiments, cannot as yet be considered entirely reliable. However, the fact of a decrease in the hydrolysis of starch in the course of perfusion is typical of those types

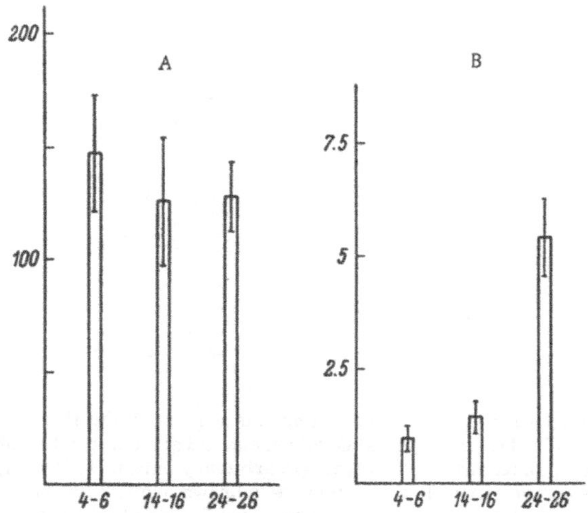

Fig. 64. Relationship of cavital and membrane digestion in early ontogenesis. (A) Hydrolysis in vivo; (B) in vitro. Abscissa: age (in days); ordinate: amylolytic activity (in arbitrary units).

of membrane digestion which are performed by enzymes which are weakly adsorbed on the surface of the intestine.

We could not observe a statistically significant difference in the rate of hydrolysis of starch in animals of 4-6, 14-16, and 24-26 days of postnatal development, although cavital digestion in the younger adult group was practically absent and in the older group it was definitely present.

The rate of spontaneous secretion of amylase (in rabbits it is continuous) is sharply increased during the period of transition from a milk to a vegetable diet, which is generally considered as an adaptation phenomenon.

It should be noted that our data regarding the entry of pancreatic amylase into the intestine and, consequently, the level of cavital digestion are in agreement with the results of other investigators. Indeed, the fact of rather late ontogenetic development of the secretion of enzymes which cleave polysaccharides has been established by all authors who studied this problem (Kruchkova, 1939; Smith, 1959; Stepanova, 1960; Kuznetsova, 1963).

On the basis of studies of only the cavital digestion one would come to the conclusion that the hydrolysis of polysaccharides in the digestive tract of the newborn is practically impossible. In reality, however, as has been shown above, hydrolysis in vivo definitely occurs even during the first few days of life and its intensity changes little during the first month of postembryonic development.

It follows that the cleavage of starch during the first few days of life is performed by enzymes which act on the outer surface of the cell membranes. This assumption is in accord with the fact that during the period of milk nourishment the main digestive glands are functioning poorly as well as with the fact that the highly dispersed products, such as milk, do not require any preliminary treatment to which they are subjected in the cavity of the gastrointestinal tract.

The experiments with soluble starch are of model character. This substrate, although it is not typical for the period of milk nourishment, was selected because it does not penetrate the cell membranes and, therefore, permits a clear differentiation between cavital and membrane digestion and excluding intracellular digestion.

Closely similar data were obtained by Toropova and collaborators during their studies of hydrolysis of starch in rat intestine and determination of the enzymatic activity in the extracts of the pancreas and in blood. Moreover, during the studies, under similar methodological conditions, of intestinal digestion of starch in rabbit embryos, Toropova observed that the intestinal surface begins to perform its enzymatic functions toward the end of embryonic development (Toropova, 1964; Toropova et al., 1965a, 1965b).

Later on in our laboratory we investigated the role of the surface of the small intestine in the hydrolysis of certain disaccharides, dipeptides, and fats employing rats of various ages (Iezuitova et al., 1964c, 1965b). The experiments were performed on inverted slices of the small intestine of rats of Wistar strain on the 10th, 20-22nd, and 31-33rd day of postnatal development and of the adults. The activity was expressed in arbitrary units per cm^2 of the surface of the small intestine.

Invertase. As can be seen from Fig. 65, invertase could not be detected in any of the 10-day-old rats. Of the five 20- to 30-day-old rats only three showed the presence of invertase and in the other two it was practically absent. It is of interest to note that the weight of the latter two rats was 1-2 grams lower than that of the rats which showed the presence of invertase activity. Even at this time there is a definite difference in the distribution of this enzyme along the intestine. The invertase activity is practically absent in the distal third of the intestine. By the 30th day the activity of invertase noticeably rises in all segments of the small intestine.

Comparison of the hydrolysis of sucrose in the presence of the intestine and under the influence of enzymes which pass into solution

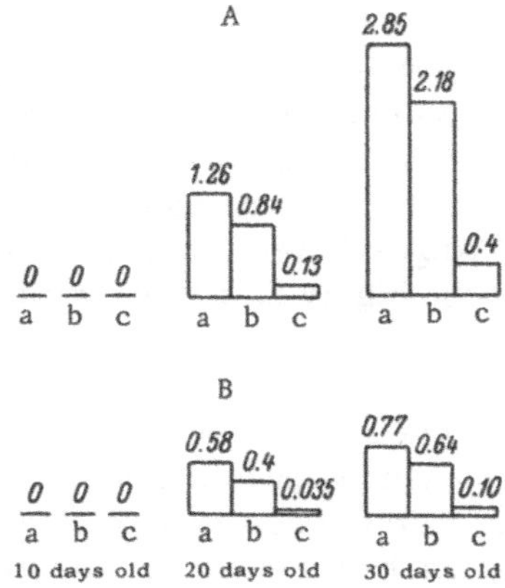

Fig. 65. Invertase activity per cm² of the surface of the small intestine of various ages (in mg % of formed hexoses per minute of incubation; average data). (A) Incubation in the presence of intestine (cavital and membrane digestion); (B) hydrolysis under the influence of enzymes passing in solution (cavital digestion). (a) Proximal, (b) middle, and (c) distal segments of the small intestine.

leads to the conclusion that the cleavage of this disaccharide is mainly accomplished by the enzymes fixed on the surface of the intestine. The activity of invertase per unit intestinal surface rises noticeably with age during the investigated period of ontogenesis.

Peptidase (Fig. 66). Since the cleavage of only one dipeptide was investigated the results obtained cannot be considered of general nature. In contrast to the results obtained above, the activity of dipeptidase was already quite high in 10-day-old animals (of the four rats only one did not show any cleavage of glycyl-L-leucine). The activity of the intestinal surface rises toward the 20th day and changes little by the 30th day.

In animals 20 and 30 days old there was no difference in the peptidase activity in the proximal, distal, and middle portions of the intestine. The impression is gained that in 10-day-old animals the

Fig. 66. Peptidase activity per cm² of the surface of the small intestine of rats of various ages (in % of hydrolyzed substrate; average data). Notation the same as in Fig. 65.

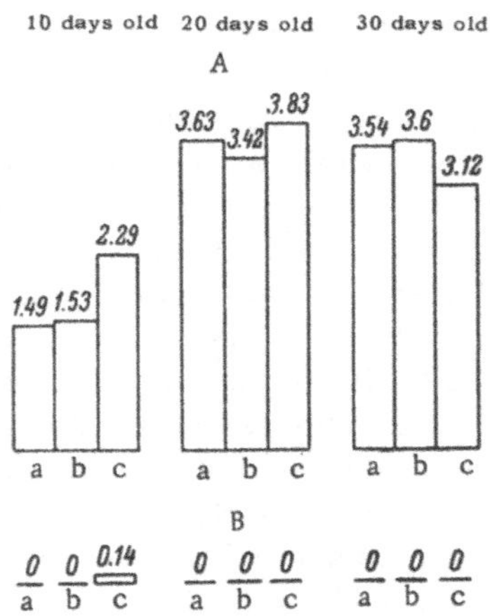

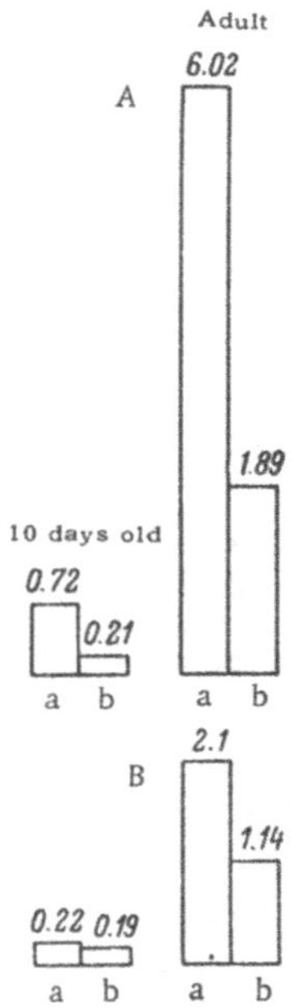

Fig. 67. Lipase activity per cm² of surface of the small intestine of 10-day-old and adult rats (in % of hydrolyzed tributyrine). (a) Proximal and (b) distal section of the small intestine. The rest of the notation is the same as in Fig. 65.

activity of the distal third is higher than in the proximal and middle sections. However, this requires further verification. The transfer of the peptidase into the incubation solution was very insignificant.

Lipase (tributyrinase). The lipolytic activity in 10-day-old rats was compared with that in the adults. As can be seen from Fig. 67, the hydrolysis of tributyrine in the proximal section in 10-day-old animals is more intense than in the distal one. The activity of the enzyme per unit surface of the intestine of adult animals is several times higher than in the 10-day-old animals. The proximal–distal gradient is also sharply defined in the adult animals.

Under the conditions of our experiments the hydrolysis of tributyrine in the proximal sections is accomplished mainly by enzymes adsorbed on the surface, while in distal sections it is accomplished almost exclusively by the enzymes which pass into the incubation medium. In adult animals, under identical conditions, the significance of enzymes acting in the cavity rises (on the basis of increase in total activity), while the role of membrane digestion in the distal sections is considerably lower than in the proximal ones.

Thus, these experiments furnish a sufficient basis for believing that membrane digestion plays a very important role during the early stages of ontogenesis. A series of facts proving that the digestion of

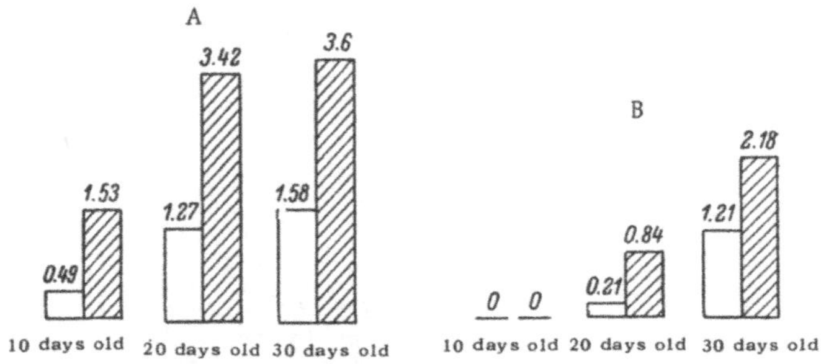

Fig. 68. Effect of mixing on the rate of hydrolysis of sucrose and glycyl-DL-leucine by inverted sections of the small intestine of rats of various ages. (A) Peptidase; (B) invertase. Light columns: without mixing; shaded columns: with mixing.

dipeptides and disaccharides is accomplished on the outer surface of the intestinal cells, and not intracellularly, was cited in Chapter II.

In this series of experiments we studied the effect of the movement of fluid in relation to the surface of the small intestine on hydrolysis of disaccharides and dipeptides. In young rats as well as in adults (Fig. 68), mixing leads to an increase by 2- to 3-fold in the rate of hydrolysis compared to controls in which the inverted sections of the intestine remained stationary.

These data serve as further confirmation of the conclusion that hydrolysis takes place extracellularly.

Experiments with repeated incubation of the inverted section of the intestine permitted Nurks to observe the desorption of tributyrinase, a phenomenon which is also characteristic of those cases in which membrane digestion is accomplished by enzymes adsorbed from the chyme.

Earlier, various facts were cited which indicated that the period of milk feeding (from the standpoint of the gastroenterologist) may be considered as a period during which the hydrolysis in the cavity of the gastrointestinal tract and the secretory function of the main digestive glands is only weakly developed, and in which digestion is basically accomplished on the surface of the intestinal epithelium. The comparatively late development of cavital digestion in prematurely born mammals was well known before the first information regarding the existence of membrane digestion appeared. The interrelationships between cavital and membrane digestion at various stages of ontogenesis are, in the final analysis, adaptations of the organism to the peculiarities of nutrition.

Cavital digestion develops to that degree needed for the preliminary treatment of the foodstuffs occurring in the cavity. It is clear that there is comparatively little of this during the period of milk nourishment and then cavital digestion sharply increases in volume and significance during the transition to definitive nutrition. This transition is sometimes difficult, and in the process of evolution sometimes remarkable means are selected to accomplish it. For instance, it is known that in many predatory animals, such as wolves, the transition from milk nourishment to meat diet has an interesting intermediate stage during which the parents feed the young regurgitated food. In this manner the young animals receive not only partially digested food but a certain amount of enzymes as well.

All of this recalls the thought expressed by Barcroft (1937) that the stage is prepared before the spectacle begins. What we know about the physiology of digestion leads one to consider at least two circumstances which supplement this truly remarkable thought which characterizes the preventive adaptive development to physiological functions: (1) the stage is prepared for the spectacle which will be played; and (2) occasionally the course of the spectacle is altered depending on how possible it was to prepare the stage.

The following example may serve as an illustration of the first circumstance.

It is well known that guinea pigs, in contrast to other laboratory mammals, are born completely mature. As has been shown in our laboratory by Stepanova (1966), the development of the amylolytic function of the pancreas in these animals is basically complete at the time of birth. On the other hand, the activity of β-galactosidase is very low during the early period of postnatal development of these animals (De Groot and Hoogendoorn, 1957; Koldovskii et al., 1966).

Thus, in guinea pigs at the moment of birth the character of digestion approaches that which, in other mammals, is observed in later periods of ontogenesis (here we speak also about the interrelationship of cavital and membrane digestion, and about the interrelationship of carbohydrases which perform the cleavage of lactose and other disaccharides present in the definitive diet).

The adaptive character of these peculiarities of digestion in guinea pigs is quite obvious.

As regards the second circumstance (changes in the course of the spectacle depending on how possible it was to prepare the stage) predatory animals feeding semidigested food to their young is an illustration of such an improvisation which arose in the process of evolution of the animal world.

Unfortunately, for a long time the physicians during the selection of a children's diet were not inclined to adapt themselves to the peculiarities of the stage and played the same spectacle over and over again. We speak here about a considerable and individual variation in the length of time of development of the enzymatic activity and of the changes in the relative role of cavital and membrane digestion. But before we speak about the errors of the physicians it is essential to dwell in greater detail on the problem itself.

It is known that during early ontogenesis the characteristics of the activity of the gastrointestinal tract are drastically altered. This is true not only for cavital but also for membrane digestion. The assortment of enzymes acting on the surface of the small intestine changes, the proximal–distal gradient of absorption and hydrolysis changes, and the activity of the enzymes changes in either direction (Bailey et al., 1956; Hill, 1956; Mosinger et al., 1959; Alvarez and Sas, 1961; Doell and Kretchmer, 1962, 1963) (Fig. 69). The transition from milk to definitive nutrition is of critical character. Thus, in rats from the 15th to the 20th day β-galactosidase activity falls from maximal to minimal values, and at the same time enzymes appear which perform the final stages of the hydrolysis of polysaccharides which are typical of mixed diets. We have noted, however, that in certain animals the development of the activity of invertase was

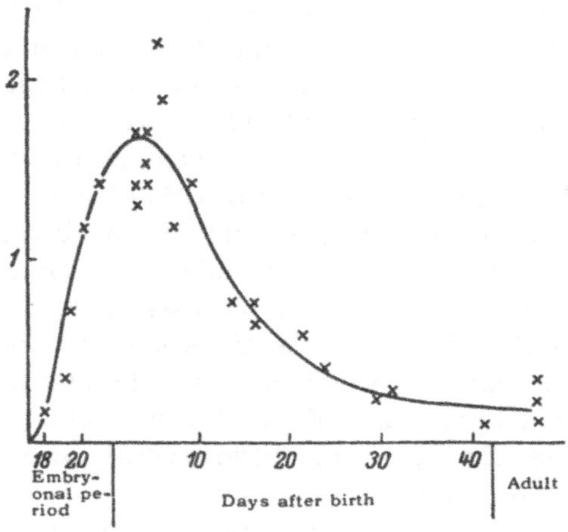

Fig. 69. Activity of β-galactosidase (substrate-lactose) in homogenates of the intestine of growing rats (according to Doell and Kretchmer, 1962, page 359, Fig. 6). Abscissa: age; ordinate: activity (in micromoles of glucose/mg protein/10 minutes).

sharply delayed. These animals fell behind the others by other criteria of development (Iezuitova et al., 1964c, 1965b).

At the present time the mechanisms which ensure the induction of the synthesis of constitutional enzymes are not well understood, that is, the mechanisms which in the final analysis determine the time of "reading" the genetic information. There are data which indicate that such processes may be induced by the hormones of the hypophysis and adrenals; thus, Doell and Kretchmer (1964) observed that on administration of corticosterone to 3- to 9-day-old rats the production of invertase begins. There was no repression of the β-galactosidase activity. Apparently, the mechanisms of the synthesis of invertase and the inhibition of synthesis of β-galactosidase are not identical. Therefore, cases are theoretically possible in which the activity of β-galactosidase will be already suppressed while the activity of invertase is not yet developed, as well as cases in which both activities are already evident. Since the induction of synthesis of definite digestive enzymes is elicited by hormones of a wide spectrum of activity, the same factors are responsible for the necessary correlation between the development of various attributes in ontogenesis. At the same time, it indicates that the delay in development may be aggravated by the enzymatic defects of the digestive system and by the subsequent alterations in nutrition. In this sense we turned our attention to the necessity of individual adaptation of nutrition to the potentialities of the digestive system of a child.

The fact that the induction of enzymes is elicited not so much by the aid of food as by the aid of endogenous factors is not a refutation of the adaptive character of such induction. The gist of the matter consists in the fact that the adaptation to the quality of nutrition acquired in the course of evolution is an inborn character and became correlated with the "requirements" of the organism, thus offering a certain degree of freedom in the selection of the foodstuffs (Ugolev, 1961a). The work of Sunshine and Kretchmer (1964) indicates the adaptability of β-galactosidase to the presence of lactose in milk. The authors based their observations on previous data of Pilson and Kelly (1962) that lactose is absent in the milk of California sea lions. In addition, the normal food of these carnivorous animals contains very small amounts of carbohydrates. Studying the young sea lions (6 and more weeks of postnatal development), Sunshine and Kretchmer showed that the administration of monosaccharides to these animals is accompanied by a rise in blood-sugar levels. After the administration of lactose the curve of blood-sugar levels was flat; a similar result was obtained after the administration of sucrose. It is of interest that it was possible to obtain in these animals not only the proof for the adaptability of oligosaccharidases but also to reproduce diarrhea which resembled that which is observed in man during lactose and invertase deficiency after the administration of the corresponding disaccharides. The subsequent study of homogenates of the mucosa confirmed the absence of lactase, invertase, trehalase, and cellobiase, and a very low level of maltase in the sea lions.

The comparative characteristic of enzymes in various species of animals is shown in Table 11.

Thus, there is a twofold mechanism which ensures the adaptability of membrane digestion to the quality of nourishment. This is the substrate–enzyme induction which we mentioned earlier, and species adaptations which are fixed by the genetic apparatus of the organism and the intestinal cells. In this case induction may be determined not by the presence of a nutrient substrate* but by a corresponding nervous or humoral stimulus (inductors, intermediaries; for details on this question see (Ugolev, 1961a).

As we mentioned earlier in the section "Topography of Membrane Digestion" (see Chapter IV), during the transition from milk to definitive nourishment not only does the assortment of enzymes which perform the final stages of hydrolysis change, but also their distribution along the intestinal tube. For instance, Ovdeichuk noted the transposition in the proximal direction of the maximum of invertase activity (Nadirova et al., 1965a), and Timofeeva noted the transposition, also in the proximal direction, of the activity of peptidase (Iezuitova et al., 1964c, 1965b; Timofeeva, 1965). In connection with this observation we expressed the assumption that such a change in the distribution of enzymes is connected with the reduction in the fraction of fat in the mixed diet as compared with that in milk.

During mixed feeding, new interrelationships are established between cavital and membrane digestion which, apparently, may alter depending on the fraction of polymers and oligomers in the ration. These questions have not been studied in greater detail.

Recently De Laey observed very important new facts which characterize the changes in membrane digestion in ontogenesis. Studying the correlation of the hydrolysis of soluble starch by amylase in the intestinal cavity and on its surface, he confirmed the previously obtained data regarding the predominance of membrane digestion in young rats. Following up the investigation of the age evolution of the

*As we mentioned earlier, the regulation of β-galactosidase activity in ontogenesis does not depend on the diet. In this connection, of great interest is the observation that during maintenance of young rats on a ration containing lactose in the same concentration as in the milk, the lowering of the activity of β-galactosidase (normally observable between the 14th and 21st day) is substantially delayed (Koldovskii and Chytil, 1965).

TABLE 11

The Activity of Intestinal Enzymes in Man, Rat, and Sea Lion (According to Sunshine and Kretchmer, 1964, p. 851)

Enzyme	Man	Rat	*Zalophus californianus*
Lactase	10.8	10.2	0
Invertase	7.2	3.0	0
β-galactosidase	1.3	3.5	0
Trehalase	–	2.8	0
Cellobiase	–	1.7	0
Alkaline phosphatase	–	12.4	3.4
Carbamylphosphate synthetase	–	14.0	7.0
Ornithine transcarbamidase	–	1372.0	314.0

correlation of cavital and membrane digestion, De Laey observed that in old rats in which, according to Baker et al. (1963), the villi are atrophied, the significance of membrane digestion is reduced to almost zero. We do not have at present a complete characterization of the correlation of cavital and membrane digestion in old animals, but the data of De Laey, in relation to the fact that in aging of the organism the processes of regeneration are altered in all tissues, permit the assumption that one of the important causes of the widespread occurrence of intestinal diseases in old age could be the alteration in membrane digestion (it is known that the normal functioning of the intestine is connected with its continuous regeneration) (see also Table 12). Thus, we are entering a field not yet known which could be called "gerontological aspects of membrane digestion."

REGARDING THE COMPARATIVE PHYSIOLOGY OF MEMBRANE DIGESTION

At the present time there is reason to assume that membrane digestion is a very widespread and, perhaps, universal mechanism ensuring the transformation of foodstuffs in multicellular animals — higher and lower —, in various monocellular organisms, and in plants. It will be remembered that hydrolysis on the outer surface of the cells is an intermediate stage between the distal digestion and absorption.

Hydrolysis by amylase, adsorbed on the surface of the isolated slices of the intestine pretreated with TCA (trichloroacetic acid), has been observed in rats, mice, rabbits, cats, pigeons, and frogs (Ugolev, 1960a, 1960b). Similar data were obtained by Lee (1962) on isolated

TABLE 12

Functional Indicators of Small Intestine in Adult and Old Animals and Man (Spencer, 1964b)

Functional indicators	Object of study	Notes
Ratio of weight of the intestine to weight of animal		
Adult age	Cat	2.8%
	Man	1.8% (including stomach; higher figures are given)
	Cow	1.8% (higher in younger animals)
	Sheep	2.1% (by 91st day; higher in younger animals)
	Rat	1.7%
	Rabbit	1.5% (weight per unit length varies in sublines)
Depending on diet	Sheep	Values may vary by 2-fold
	Rat	Higher on lactose than on starch diet
Depending on age	Rat, man	Decreases with age
Absolute weight of the intestine	Man	Probably, highest indicator at 41-50 years of age
Excretion of ascorbic acid with feces	Man	Decreases
Absorption of calcium	Rat	In vitro the absorption and transport is higher in growing rats than in old. However, data are available which indicate that absorption is higher in aged than in adult rats
Time of renewal of intestinal epithelium cells	Mouse	Greater in old mice; the transition from crypt to end of villi is longer
Incorporation of cytidine-H^3 into intestinal cells	Man	Lower in older mice
Transport of magnesium	Rat	Insignificant difference between old and young animals
Phosphatases	Rat	Activity decreases with age
Resistance to nematodes	Cattle	Adults more resistant than calves
Size of intestinal cylindrical epithelium	*Bufo melanostictus*	Cells of intestinal epithelium larger in size in large than in small animals
Absorption of thiamine	Rat	Smaller in 2-year-old rats than in young rats
Absorption of vitamin B_{12}	Man	Unchanged in aging, although concentration in blood may be decreased

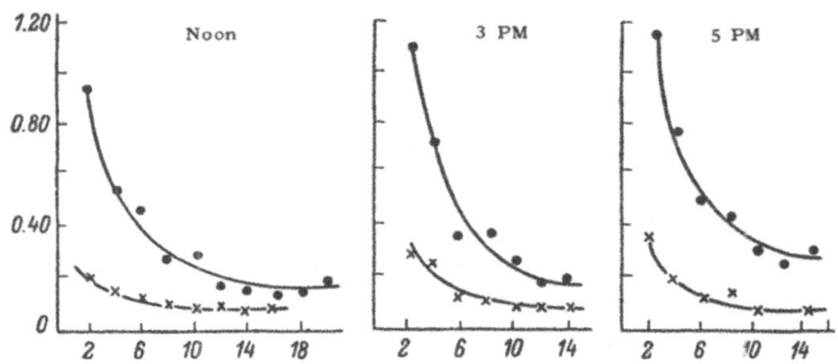

Fig. 70. The restoration of amylolytic activity of an isolated slice of the small intestine after chyme has been passed through it in a ram, named "Boy." Abscissa: time (in minutes); ordinate: hydrolysis of starch (in mg/minute). The numbers above the curves indicate the time of day during which experiments were performed.

pieces of the intestine of chickens. The membrane effects during perfusion by a solution of starch of the sections of the small intestine of rats, guinea pigs, white mice, hamsters, and frogs were noted by Ugolev (1961a, 1963). Invertase phenomena were described for rats, hamsters, and guinea pigs. The significance of the surface of the intestine in the cleavage of certain peptides has been studied only in a few mammals, e.g., rats and guinea pigs (Ugolev, 1963).

In the laboratory of Yunusov, Rakhimov (1965) and others demonstrated that the membrane hydrolysis of starch in the small intestine of sheep is very significant (Fig. 70). These first data regarding membrane digestion in ruminants are basically concerned with the role of this mechanism in the process of the transformation of food. Further investigations are required for a more complete characterization of this mechanism. For instance, the question whether membrane digestion in ruminants occurs exclusively in the small intestine (as in other mammals) or whether it is possible also in certain sectors of the stomach, where an intensive resorption of certain nutrients takes place, is of interest.

There are grounds for believing that the interrelationships between cavital and membrane digestion in birds and mammals have much in common. For instance, in chicks the contrast between a very high amylolytic and very low maltase activity in the contents of the small intestine is striking (Laws and Moore, 1963). This leads to the conclusion that in birds the hydrolysis of polysaccharides begins in the intestinal cavity and is continued and finally accomplished on the surface of the intestinal epithelium.

Quite recently Berman (1964) and Salenietse investigated the role of the surface of the small intestine of carp in the cleavage of starch.

They showed that the enzymes adsorbed from the chyme play an important role in the performance of the digestive processes in fish. The same authors were able to observe a typical amylase desorption from the surface of the small intestine. Along with these data, indicating the existence of membrane digestion in fish, Berman and Salenietse showed an increase in the enzymatic activity of the intestinal surface in year-old carp as compared to newly hatched ones. Of interest are their observations regarding the fact that the amylolytic activity of the intestinal surface is subject to seasonal fluctuations (in the summer it is considerably higher than in the winter).

The similarity in structure and ultrastructure of the intestinal surface in various vertebrates and the general plan of functioning of the small intestine leads to the conclusion that membrane digestion is present in all higher animals. The presence of microvilli, typical of the intestine, permits the postulation of this hypothesis in regard to the invertebrates as well.

Thus, in round worms (ascarid) the structure of the brush border is the same as in higher animals (Zetterqvist, 1956; Tokin, 1959). Later, Kessel et al. (1961) investigated in detail the structure of the intestinal cells of *Ascaris lumbricoides*. They showed that the free surface of the cell forms considerably taller microvilli than usual (6 microns in length and 0.1 micron in diameter). According to the calculations of these authors, the surface of the individual cell contains more than 3500 microvilli and because of the microvilli the surface of the cell increases 75-fold. For comparative physiology, the branching of the microvilli and the presence of bends or twists could be of considerable interest.

A histochemical study of aminopeptidase in *Ascaris lumbricoides* showed that in the contents of the intestine, fixed with formaldehyde, this enzyme is concentrated in two thirds of the cell adjacent to the free surface (Lee, 1962). Further investigations will, probably, furnish greater details regarding the localization of functions of this enzyme.

In the well-known review by Oppenheimer and Kuhn (1932) there are indications that the peptidases of the intestine of ascarids are firmly bound to cell structures and are desorbed with great difficulty. Thus, in a number of lower invertebrates one could assume the existence of membrane digestion either on the basis of the similarity of the ultrastructure of the intestine with that in higher vertebrates, the absence of attributes of intracellular digestion, and the presence of a clearly expressed enzymatic activity of the intestinal cells.

There are several convincing proofs of the fact that invertase is localized on the surface of yeast cells, the activity of which is essential for the utilization of glucose and fructose (Demis et al., 1954; De La Fuente and Sols, 1962; Sutton and Lampen, 1962). In special experiments Best (1955) showed that invertase is present on the outer surface of yeast membrane, while hexokinase is localized on the inner surface of the membrane. It is possible that a large number of various hydrolytic enzymes, forming a peculiar zone of membrane digestion, are localized on the outer surface of the yeast cell membrane. For instance, Myrbäck and Vasseur (1943) detected lactase on the surface of yeast cells, Myrbäck and Ortenblad (1936, 1937) detected trehalase, and Rothstein and Meier (1948) and others, phosphatase. In this case phosphatase is essential for the utilization of phosphorylated sugars which penetrate the yeast cell only after dephosphorylation.

Among the methods which made it possible to show that various oligosaccharidases are localized on the outer surface of yeast membranes and certain microorganisms are the widely used methods which are based on the sensitivity of these enzymes to high acidity of the surrounding medium. For instance, Willstätter and Lowry (1925) observed that the treatment of yeast with 0.3 *N* H_2SO_4 results in a sharp inhibition of the invertase and maltase activity. The same was observed after treatment of invertase of *Myrothecium varrucaria* and cellobiase, trehalase and maltase of *Aspergillus luchuensis* with 0.1 *N* HCl (Mandels, 1953). In all of these cases the acid of the incubation medium did not affect the metabolism and such vital functions as growth and division of the cells.

Rothstein (1954) correctly states that a comparison of these leads to the conclusions that the hydrolytic enzymes in yeast are localized on the outside of the cell membrane. It is quite significant that sucrose as such does not penetrate the yeast cells (Demis et al., 1954).

The observations of Wilkes and Palmer (1932) and Mandels (1953) regarding the effect of the pH of the medium on the activity of invertase also speak in favor of the surface localization of the invertase of yeast. Vasseur (1951) proposed that the hydrolytic enzymes of the sperm of the sea urchin are localized on the cell surface. Various phosphatases on the surface of yeast cells have been described (see review by Rothstein, 1954).

A large number of investigations also indicate the presence of phosphorylating enzymes on the outer surface of the cell membrane of various animals (Rosenberg and Vilbrandt, 1955).

In the presence of starch *Streptococcus bovis* develops an adaptive amylase (Rahn and Leet, 1949). It is assumed that the enzyme is localized on the outer surface of the membrane.

A series of investigators observed the presence of enzymatic activity on the outer surface of protoplasts of various plant cells. For instance, Brown (1955) in very clever experiments has shown that a very intensive hydrolysis of sucrose and the transformation of glucose to 2-keto-gluconic acid take place on the outer surface of the protoplast. It is of interest that because of this process a rapid growth of the isolated segments of plant roots is ensured, and, more importantly, along with the growth of the segment there occurs an increase in the activity of invertase on the surface of the protoplast. According to the author, this indicates that with growth the formation of new "enzymatic islets" takes place, between which an inert material is stored, thus dividing them from each other. By employing special conditions Brown was able to prevent the entry of sucrose inside the cells and in this manner to exclude intracellular cleavage. During this process the entry of the enzymes into the surrounding medium was not observed.

One cannot escape seeing in this example a peculiar analogy to membrane digestion in the plant world.

We have expressed the hypothesis that since membrane digestion exists in such suborganized organisms as yeast and bacteria, one cannot exclude that in the process of evolution this mechanism was not lost by various "nondigestive" cells of higher animals (Ugolev, 1963). From this standpoint the data regarding considerable phosphatase activity on the surface of erythrocytes are of interest (Clarkson and Maizels, 1952; Herbert, 1952; and others). Morrison and Neurath 1953) observed proteolytic enzymes in erythrocyte membranes. The localization of these enzymes was not established more exactly.

Ugolev and Chernousova (1963) perfused the vascular system of frogs with Ringer solution which contained soluble starch and found that the rate of hydrolysis of starch during perfusion was many times higher than could be expected from measurements of the amylolytic activity of the perfusate. Since the vascular cells are impermeable to starch, it was concluded that at least certain polymers can be hydrolyzed by enzymes bound to the structures of the small vessels. It is also of interest that the same effect is produced by the vessels of digestive organs as well as by the vessels of the lower extremities. This problem has not been investigated in higher animals.

Thus, the available data permit the belief that membrane digestion is a very widespread mechanism.

The fact that membrane digestion is present in lower as well as in higher organisms, differing in the level of organization as well as in peculiarities of nutrition, leads one to believe that there are significant differences in the course of membrane digestion in various species of animals. The facts, however, which are sufficient for the conclusion about the universality of membrane digestion are not sufficient a description of the evolution of membrane digestion. For this reason one should accept all the subsequent narrative somewhat critically.

Membrane digestion is considerably older than the apparatus of the microvilli since it is encountered in bacteria and yeast. Apparently, the smooth surface is sufficient for the exchange between a cell and the surrounding medium; but when multicellular organisms arose and there appeared specialized functional systems, such structures as the microvilli had to appear. Indeed, if one takes into account that because of the microvilli the free surface of the intestinal cell increases about 30-fold, it becomes clear that because of this fact a large number of cells are freed to perform other functions. As we mentioned earlier (Ugolev, 1963), the fact of the ultraporosity of the brush border imparts to it new properties, peculiar to very small capillaries, which are preserved through all further evolution. It is possible that the variability of the structure of the villi and, particularly, of the ultrastructure of the microvilli and of their mutual distribution in the brush border is of significance to the adaptability of the intestinal functions to various conditions of nutrition.

Furthermore, the change in ultraporosity of the brush can be observed in the same animal throughout its ontogenesis. For instance, it has been shown during the first month of life of chicks that there was an increase in the number of microvilli and their length was also increased (Overton and Shoup, 1964). Interesting changes in the ultrastructure of the brush border in connection with the metamorphosis of large frogs (*Rana catesbeiana*) were described by Bonneville (1963). This author has shown that the microvilli are present in the larva as well as in the adult specimens, although in the latter the microvilli were shorter, the ectoplasmic zone was lower, and the rough surface elements of the endoplasmic reticulum were more numerous in the supranuclear region.

While characterizing the evolution of membrane digestion in higher animals one should not limit oneself to only the functions of the villi and individual intestinal cells. The small intestine, apparently, cannot be looked upon as a very large population of intestinal cells. It is well known at present that this is a tube, whose various segments are well specialized.

In Chapter IV and early in this section we cited data which permit the characterization of the proximal–distal gradient of enzymatic activities as a peculiar device. As can be seen from Fig. 71, as a first approximation it is convenient to consider the intestine as an adsorption column (keeping in mind that the adsorbed substances are subject to further hydrolysis and absorption). Depending on the com-

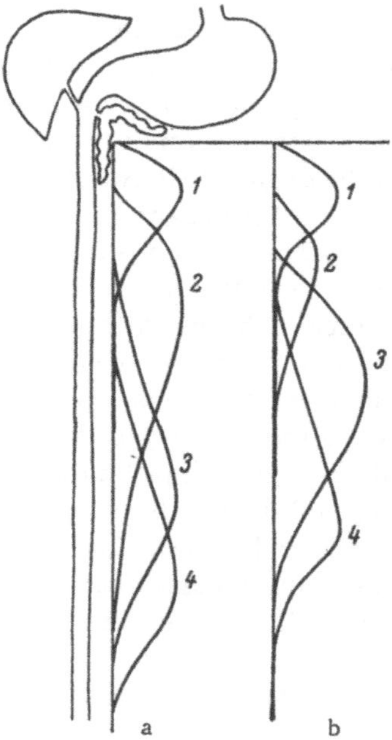

Fig. 71. Possible sequence of distribution of various substances on the surface of the small intestine depending on larger (a) or smaller (b) content of fat in the diet. (1) Bile acids; (2) fats; (3) poly- and oligosaccharides; (4) poly- and oligopeptides.

position of food the zones of adsorption of carbohydrates, proteins, and fats may become larger or smaller, and become mixed in the proximal or distal directions. For instance, with an increase in the proportion of fat the zones of hydrolysis of peptides and oligosaccharides will shift in the caudal direction. From this standpoint the length of the intestine may be looked upon as a peculiar adaptive device toward the quality of the food, to which we drew attention earlier (see reviews by Razenkov, 1948; Ugolev, 1963; Spencer, 1964a). At the present time, we can add membrane digestion to the already known factors which determine the length of the intestine.

The reduction (within certain limits) in the rate of movement of food along the small intestine increases the effectiveness of hydrolysis in each of its segments. Thus, for instance, is understood the slowdown in the motor movements of the small intestine after resection of the stomach (see review by Filippovitch, 1962). It is possible that the rate of movement is somehow correlated with the rate of transfer of the substrates from the intestinal cavity to its surface. The rapid hydrolysis of substances in the brush border presupposes the entry of newer and newer portions of the substrates which occurs either through the movement of the next portions of chyme in the caudal direction or as a result of reverse movements of the chyme.

Probably, much in the mechanism of adaptation of membrane digestion and in species peculiarities of the motor movements will become clearer after the mutual connection of these phenomena is examined experimentally. It would appear, however, that one should consider the changes in the assortment and properties of the enzymes which perform membrane digestion as the most important element in the adaptation of the intestine to the peculiarities of nutrition. It has been established (Ugolev, 1963; Iezuitova, 1965) that the ratio of invertase to amylase on the intestinal surface in guinea pigs is considerably higher than in rats.

Sunshine and Kretchmer (1964) cited interesting data regarding the ratios of various enzymes which participate in membrane digestion in adult animals (Table 11). The absence of carbohydrases in sea lions attracts attention; in species whose milk does not contain lactose, lactase is absent.*

The variation in the enzymatic composition of the intestinal surface is a phenomenon which is well known at present. In many cases it is possible to prove that a connection exists between the composition of the enzymatic layer and the qualitative aspects of the food. The enzymatic composition of the intestinal surface may also change as a result of the species differences in the adsorption properties of the cell membranes. For instance, amylase in guinea pigs is fixed on the surface of the intestine considerably more firmly than the amylase on the intestine of rats (Ugolev, 1963; Iezuitova, 1965).

Refined species peculiarities of the enzymes which perform membrane digestion, the significance of which is not always clear, may be illustrated by the following example. Apparently, there are two β-galactosidases, one of which has an optimum pH of 3.5, the other, of 5.8 (De Groot and Hoogendoorn, 1957; Doell and Kretchmer, 1962; Koldovskii et al., 1966). It is of interest that while in rats, guinea pigs, and mice the maximum activity is observed at pH 3.5, in rabbits it is at pH 5.8.

For the characterization of the evolutionary aspects of membrane digestion we must once more recall the data cited in Chapter III. In this chapter it was shown that the enzymes included in the composition of the cell membranes possess properties which substantially differentiate them from those possessed by the same enzymes in solution. In this connection, an immediate question arises: should the evolution of membrane digestion be looked upon as an evolution of properties

*Koldovskii kindly informed us of his experiments which he conducted with Dr. Sunshine on sea lions whose milk contained lactose. In the intestine of these animals the corresponding enzyme was present.

of individual enzymes or as an evolution of enzyme–membrane complexes? In all probability, the selection of useful attributes proceeded along the second pathway, since the actual activity of enzymes which perform membrane digestion takes place only when they are component members of the membrane. It is not surprising, therefore, that invertase, for instance, separated from the membranes possesses high activity only within a very narrow range of pH. It could also be assumed that many enzymes, active as components of the membrane, lose their activity or change it sharply after desorption.

But, of course, the evolution of membrane digestion is not an evolution of individual enzymatic activities. Apparently, of no lesser significance could be the evolution of the enzymatic ensembles, a question which up to the present has not been investigated. Yet, it is possible that the alteration in spatial distribution of individual enzymes, enzymes and pores, along which the resorption of the products of hydrolysis takes place, may produce a very large biological effect.

Finally, one should keep in mind the alteration in the interrelationships between cavital and membrane digestion which depend on the properties of food, as has been shown in Chapter IV and in the preceding section of the present chapter.

Earlier we emphasized on several occasions that in the majority of higher animals membrane digestion is a function of the small intestine. There is no reason to believe, however, that this rule always prevails. It would be more proper to state that membrane digestion may be performed in those organs which do not contain very large polymers of foodstuffs and in which absorption takes place at the same time. From this standpoint the data of Izmailov (1965) regarding the possibility of membrane digestion in the forestomachs of ruminants are of great interest.

If one examines this question even further one would find that membrane digestion is performed in multicellular animals not only in the gastrointestinal tract and not only by the elements of entodermal origin. We earlier mentioned the hydrolysis of starch in vessels. Earlier (Ugolev, 1963), the possibility was discussed that membrane digestion in *Pogonophora* is accomplished by tentacles.

In the study of the ultrastructure and the enzymatic activity of the surface of certain parasites, Ginetsinskaya et al. (1966) showed that in two species of sporocyst the ectodermal surface is covered not by cuticle, as was thought earlier, but by a typical brush border which contains alkaline phosphatase.

Thus, in cases where the differentiated gastrointestinal tract is absent, membrane digestion may be performed by structures which form the outer surface of the organism and which acquire considerable resemblance to the surface of the intestine.

We do not think that this exhausts all the possible paths along which the evolution and the adaptation of membrane digestion among living creatures proceeded; perhaps, in the not so distant future, other paths will become known. Something else is of importance. Membrane digestion, being the most perfect mechanism which unites the processes of cleavage and utilization of foodstuffs, varies from species to species, and as a result, probably, its maximal (for the corresponding conditions) effectiveness is assured.

THE ORIGIN OF DIGESTION – THE ORIGIN OF MEMBRANE DIGESTION

It was only recently (Florkin, 1947; Koshtoyants, 1950; Livanov, 1955; Buddenbrock, 1956; Barrington, 1962) that the evolution of digestion appeared to be a simple and unalterable scheme. The digestion of higher animals, characterized by the development of secretory processes, arose from the intracellular digestion of lower animals. In the course of this process the transition took place somewhere at the stage of evolution, higher than that of the sponge in which the digestion, according to Yonge (1937), is exclusively intracellular. No matter how enticing this simple scheme is, we must refuse to accept it.

Earlier we cited a series of facts (Ugolev, 1961b) which prove that the intracellular digestion is no older than secretory digestion. In fact, all phagocytic forms are capable of secretion of a morphostatic type (not connected with the change of cell structure and which is characteristic of higher animals). Moreover, the most primitive organisms known at present, bacteria, are also capable of true secretion, although they do not possess an apparatus which ensures phagocytosis and pinocytosis.

There is also no basis whatsoever for considering intracellular digestion more ancient than membrane digestion. The phenomena which in principle are similar to membrane digestion (hydrolysis of foodstuffs by enzymes adsorbed on the membrane surface) can be observed in such suborganized organisms as yeast and bacteria.

Thus, the facts available to date do not offer the possibility of considering that the extracellular digestion arose in the course of phylogenesis later than intracellular digestion or from it. Moreover, it appears more likely that it is the intracellular digestion that is less ancient than the other two forms of digestion.

Naturally, a question then arises as to how did the digestive processes arise in the course of evolution.

We assume that the basis for the development of various forms of digestion, that is, the sum total of the processes which ensure the hydrolysis of polymeric molecules, were enzymatic systems which ensured the mobilization of fat, protein, and carbohydrate depots of ancient organisms. But then one must assume that the intracellular hydrolases arose earlier than any other forms of digestion.

Until recently such a posing of the problem appeared to be ill founded. Indeed, it was thought that the earliest living forms were the autotrophic animals. However, lately this point of view was subjected to an annihilating criticism. Autotrophic organisms must possess very complicated systems which should ensure chemosynthesis and photosynthesis of complicated organic substances. It is very likely that the most ancient creatures did not possess a very complicated inner organization, and autotrophy arose comparatively late. The most ancient organisms, however, obtained organic substances in the form of monomers from the surrounding medium and condensed them, thus transforming them into the structures of their own organism.

Thus, the most ancient organisms were heterotrophic animals (the name is an arbitrary one). This point of view is in accord with more recent data which are concerned with the origin of life (Bernal, 1959; Hoffman-Ostenhoff, 1959; Obel, 1959; Oparin, 1959; Pirie, 1959; Fox, 1959; see also Symposium "The Origin of Prebiological Systems," 1966). Under favorable conditions even at this stage an excess of organic materials could have accumulated as peculiar depots. It is possible that these depots were not sharply differentiated but were some kind of structures which were utilized during the inadequate inflow of organic material from the surrounding medium. The hydrolytic enzymes, the most ancient of enzymes known to us, served for the mobilization of the depots.

Nutrition by complex polymeric substances of organic origin which required the presence of digestive processes and which may be characterized as the secondary heterotrophism arose considerably later. Thus, digestion arose considerably later than did the hydrolytic enzymes in general.

In what manner could the secretory processes arise?

As far as we know, until recently there was only one theory which explained the origin of cavital digestion in higher animals, the theory of Jordan.* This investigator, as well as all his contemporaries, con-

*For a detailed criticism of the theory of Jordan see the monograph (Ugolev, 1961b).

sidered that the digestion in higher animals arose from intracellular digestion in the course of evolution.

Jordan assumed that the release of enzymes into the lumen of the digestive tube without the destruction of the structure of the cells is a comparatively late phylogenetic acquisition. The initial digestion in the intestinal cavity took place exclusively as a result of the fact that during natural dying out of the endodermal cells covering the mucosa of the intestine, and during the subsequent decomposition, a certain amount of the intracellular enzymes entered the lumen of the intestine. This form of digestion, during which the enzymes enter the lumen of the digestive tube during the destruction of the intestinal cells, was, according to Jordan, more beneficial for the organism than intracellular digestion. Jordan called this form of secretion morphokinetic, since the release of the digestive secretions was connected with the alteration in cell structure.*

Therefore, according to Jordan, the most ancient type of digestive juices was the protoplasm of the discarded cells.

Later, the release of the enzymes into the lumen of the gastrointestinal tract was connected not with the death of the whole cell but only with the discarding of the apical segment of the protoplasm.

According to Jordan, after each secretory act a regeneration of the lost portion of the cell could take place. It is this type of secretion that is morphokinetic. Further evolution of the secretory apparatus is connected with the reduction in the alteration of the structure of the cell in the process of secretion. One can imagine that sooner or later a type of cell can arise, whose secretory activity takes place without any significant change in cell structure. And this is already a morphostatic secretion which, as we mentioned earlier, is characteristic of the higher animals.

The theory of Jordan appears obsolete because it connects the appearance of secretion with higher stages of evolution, whereas numerous investigations are available at present proving the presence of true secretions in very primitive organisms, including bacteria.

Then how can we visualize the origin of digestion which occurs as a result of secretion of enzymes beyond the limits of the cell?

We have proposed an excretory theory of the origin of secretory processes, the gist of which consists of the following. We began with the fact that the entry of various substances into a living system during the process of assimilation and the exit from it of products of dissimilation may be considered as ancient as life itself.

*We suggest calling this form of secretion morphonecrotic (Ugolev, 1961b) since the entry of enzymes into the intestinal cavity is connected with the death of the cells.

It would be very rash to think that the organization of the cells even in the highest forms is so perfect that during excretion only the final, unneeded products of life activity are eliminated.

The best proof of the incorrectness of this assumption is the presence in the urine of mammals of a variety of enzymes, including the digestive ones—amylase, pepsinogen, and trypsinogen. It is hardly necessary to prove that the presence of the latter in the urine has any physiological significance, and it only demonstrates that during excretion, which is performed by such highly differentiated structures as the kidneys of mammals, a certain number of biologically active substances is eliminated. It is probably true that in the organism there is not a single cell which, while eliminating mainly the products of decomposition, was not losing a certain amount of biologically valuable substances. And if we still tend to talk about the perfection of the organization, it is only because in the process of excretion the elimination of the injurious products sharply predominates over the loss of the useful ones.

If we assume that the elimination function was perfected in the process of evolution, then it should follow that in the primitive forms, from which one must begin the evolution of the extracellular digestion, the loss of intracellular enzymes during excretion was considerably greater. In freely living *Protozoa* the elimination of enzymes one must consider as an inevitable loss which had to diminish in the process of evolution. On the contrary, in those representative of this type, which began adapting to a parasitic way of life, the hydrolytic enzymes on elimination found appropriate substrates. The products which arose in the course of activity of these enzymes outside the cell became an additional source of nourishment. The elimination of hydrolases became a useful function which became gradually stronger and was perfected. Many microorganisms and *Protozoa* eliminate various hydrolytic enzymes into the medium in which they grow (for instance, proteases and carbohydrases). In many cases it was possible to show that here we are dealing with a process which is similar to the secretion of a morphostatic type in higher animals (see review of literature: Stefenson, 1951).

Returning to the multicellular animals, it should be mentioned that excretion into the closed cavity of the intestine, after securing the prey, was quite useful even in the presence of a small amount of enzymes. Indeed, the accumulation of products of decomposition in the intestinal cavity should produce injury to the cellular structures of the food object and, consequently, lead to faster death and autodigestion of the prey. This was facilitated by the enzymes which were inevitably eliminated during excretion. With the course of further evolution the increased content of the hydrolytic enzymes in the excreta required in the digestion was stabilized, while the loss of other enzymes was either kept at the same low level or even decreased. Gradually the excretion was transformed into a digestive secretion. It would appear that the excretory hypothesis of the origin of secretory processes may explain the presence in the composition of digestive juices of substances such as the group and type antigens, enzymes which ensure respiration and oxidative phosphorylation, etc.

One unexpected circumstance must be emphasized. The digestion of the heterotrophic animals is an initial stage of assimilation, while secretion is the final link of the dissimilatory processes. If our assumption is correct, then it should follow that the external secretion, that is, an essentially assimilatory process, must have arisen on the basis of a typically dissimilatory process.

Thus, in contrast to Jordan, we assume that the morphostatic secretion arose not from the morphokinetic secretion but from the morphostatic excretion.

The cavital digestion of higher animals connected with secretion arose in this manner.

As regards membrane digestion, the characterization of its origin is difficult. We can only make a few suppositions.

The enzymes, which in the process of excretion were eliminated to the surface of the cell, could be adsorbed at this site, thus forming an enzymatic layer whose effectiveness increased with evolution. On the other hand, it could be assumed that the processes of enzymatic cleavage on the surface of living cells arose even before the membranes existed as definite structures. The enzymes of the protoplasm, which happened to be inside the latter, performed the mobilization of the depot, while the hydrolytic enzymes on the surface of the protoplasm performed the hydrolysis of the polymeric molecules of the solution which surrounded such organisms.

In any case it is clear that intracellular digestion, extracellular membrane digestion, and extracellular distal digestion (connected with secretion into a special cavity or into the surrounding medium) arose on the basis of enzymatic systems which ensured metabolism in living structures.

In the course of these discussions of the origin and evolution of various forms of digestion, in several instances we made an error which could be called traditional. The fact is that almost invariably we were examining interrelationships and possible intertransitions between forms of distal, membrane, and intracellular types of digestion existing at present. At the same time it is clear that each one of these processes is a result of a prolonged and complex evolution.

It would have been more correct, therefore, to examine the most ancient original forms of digestion. Earlier we attempted to present a picture of how hydrolases, which originally performed the function of mobilization of a depot, gradually began performing the functions of hydrolysis of dietary polymers present in the surrounding medium. The diffusion of the enzymes into the surrounding medium could be looked upon as the most primitive distal digestion. The hydrolysis of the substrates which were diffused inside the protoplasm was, possibly, the original intracellular digestion and the cleavage of substances which were in contact with the surface, the original membrane digestion. Depending on several circumstances, each of these types of digestion could become dominant.

As we have mentioned in Chapter I, the hydrolysis of large molecules is effectively accomplished mainly by the enzymes in liquid phase, and this circumstance could have been an important factor in the development of distal digestion. On the contrary, nourishment by small molecules should favor the membrane and intracellular digestion.

Here we should mention a circumstance which, apparently, was responsible for the disappearance of the original intracellular digestion. We have in mind the development of cell membranes, whose presence on the one hand, is a necessary condition for the origin of a whole series of systems which regulate the intracellular metabolism and homeostasis and which, on the other hand, materially prevent the entry of various substances, particularly the poly- and oligomers inside the cells. Indeed, in bacteria and yeast the processes of digestion are either distal or membrane.

Contemporary forms of intracellular digestion are incomparably more complex. From the standpoint of the course of the process itself, conditions of contact between the foodstuffs and enzymes which cleave them, these should be considered as a miniature variant of distal or membrane type of digestion within the territory of the cell. The intracellular digestion in organisms living today is always performed extraplasmically, in special vacuoles surrounded by a membrane which either resembles or is identical to the cell membrane. The enzymes act either in solution (and then it is a microcavital digestion) or on the inner surface of the membrane (and then it is a membrane digestion). But, in addition, intracellular digestion always assumes the development of a special apparatus, which ensures the formation of digestive vacuoles, and this makes it more complex than the distal and membrane digestion. Yet, in the classical literature the intracellular digestion is generally considered as the most primitive of all the existing forms of digestion.

It could, therefore, be assumed that contemporary intracellular digestion arose on the foundation of the distal and membrane digestion as a result of increasing complexity of these types of hydrolysis. Such increasing complexity could have been useful when the digestion of large molecules or food particles could not be accomplished beyond the limits of the cell. Such a situation exists, for instance, for the monocellular organisms in very large volumes of fluid, when the secretion into the surrounding medium is useless because of dilution of the secreted enzymes, and the membrane digestion is ineffective because of the unavailability of the hydrolyzable bonds of the large molecules. From this standpoint, there are two more factors of no lesser importance. The intracellular digestion is not only a method for the transformation of food, as we mentioned earlier, but also a method for securing it, in the first place, and its deposition, in the second. From this standpoint, phagocytosis and pinocytosis are not only digestive but also food-procuring processes.

The expressed hypothesis is a first attempt to characterize the origin and evolution of digestive processes, keeping in mind the existence of three basic types of digestion (and not two, as has been assumed heretofore). Our hypothesis is based on rather shaky ground. What we have in mind is the brevity of information regarding the phylogenesis of digestive functions and, to a greater extent, the extremely contradictory nature of modern concepts regarding the origin of life.

Conclusion

Now that the book is finished we consider it important to emphasize once more that it cannot claim the completeness of illumination of the physiology and pathology of membrane digestion. Nevertheless, we hope that the reader will be able to make at least three deductions.

1. There are not two, as has been previously thought, but three basic types of digestion: distal, intracellular, and membrane. All three mechanisms, ensuring the transformation of foodstuffs, are encountered at various steps of the evolutionary ladder, in the majority of cases as combinations and while interacting with one another.

Apparently, at the present time the characterization of digestive functions of a given species of organism cannot be accepted as sufficient if it does not include the necessary information regarding the role of distal, intracellular, and membrane digestion, although, of course, the significance of each could vary within wide limits.

2. The classic scheme for the work of the gastrointestinal tract of higher animals and man, which included two phases, distal digestion and absorption, is completely insufficient and it should be replaced by another which includes three equally important phases: distal digestion, membrane digestion, and absorption.

3. It is essential to sharply distinguish membrane digestion as a most important mechanism essential to life, which connects in one continuous chain the initial processes of hydrolysis in digestive cavities, the processes of absorption in intestinal cells and the first attempts (and for this reason imperfect) to characterize this mechanism.

Although today the theoretical and applied importance of the investigations of membrane digestion is evident, one should remember that this work is only beginning, and that many problems are awaiting their investigators.

Appendix

Supplementary Information*

Unfortunately, within the scope of a brief appendix it is possible to cite only some of the many new data regarding membrane digestion in bacteria and yeast and the interrelationship of the hydrolytic and transport processes that have been published since the Russian edition of this book was printed. It should be noted that in a detailed work of Kaplan and Tacreiter (1966) it was demonstrated that organic and inorganic derivatives of mercury, which do not penetrate the cell, inhibit β-galactosidase while little affecting other enzymes which are distributed on the inner surface of the cytoplasmic membrane. It is of interest that the precedence of hydrolysis over transport across the membrane, according to the authors, is just as essential in this yeast mutant as it is in higher animals, since disaccharides cannot be transported across the membrane.

In the course of the presentation in this book we drew attention repeatedly to the fact that membrane digestion is not only an important mechanism which ensures nutrition, but that it acts as a peculiar enzymatic barrier which prevents the entry of foreign substances into the intracellular medium. In this connection of great interest is the observation of Garber and Citri (1966) that in the culture of *Bacillus cereus*, adapted to penicillin, penicillinase is localized on the surface of the membrane.

In Chapter V we cited data which demonstrated that a comparison of the activities of intact pieces of the small intestine and their homogenates in the same experiment is of great interest in that it permits a comparison of the total depot of various enzymes in the intestinal cells with the quantity of the same enzymes localized on the surface and which perform the membrane digestion. This methodological approach allowed the detection of a whole series of correlations. Thus, on comparison of these indicators of the activity of invertase and maltase of the intact and homogenized mucosa of the small intestine of rats sub-

*This section was prepared specially for the English edition.

jected to lead poisoning, Nadirova and Ugolev (1967) observed that while the hydrolysis of maltose by the intact pieces of the intestine decreased by more than twofold, the activity of the homogenates not only did not decrease but increased. The activity of invertase in the intact cells of poisoned animals was the same as that in control animals, while in homogenates it increased. It follows from these experiments that the effect of damaging factors on various disaccharidases of the intestinal cells is not the same. While the total maltase activity, determined by several enzymes, decreases there is an increase in the activity of one of the maltases, namely, invertomaltase, which, perhaps, should be looked upon as a compensatory reaction within the limits of one enzymatic system. The mechanism of such compensatory reaction is possibly connected with the peculiarities of the synthesis and translocation of various maltases. An increase in synthesis, in the course of insufficient maltase activity of the surface, may be a result of the fact that the state of the system is controlled by its end effect which depends not so much on the synthesis of enzymes as on their activity on the membranal surface.

Another important fact was observed by Gavrilov and Damanskii (1966) who studied the dynamics of the alteration and restoration of membrane digestion during ionizing radiation of rats (in doses of 450 to 700 R). These authors confirmed the previously observed data (see Chapter V) regarding a sharp decrease in the synthesis of invertase on the 3rd day following irradiation, and described the dynamics of the restoration of membrane digestion during the interval of 4 to 30 days. It is of interest that in late periods there was a prolonged and unmistakable decrease in the activity of homogenates while the activity of intact intestinal cells remained normal. Apparently, in this case the processes of translocation of the synthesized enzymes are not only not weakened, as in the case of the lead effect, but increased, which the authors regard as a peculiar compensatory reaction.

Judging by many external symptoms such an important disease as dyspepsia of newborn calves resembles certain alterations which arise as a result of defects in membrane digestion and absorption in man and laboratory animals. Of considerable interest are the first observations made by Shishkov (1966). Histochemical investigation of the small intestine of affected calves has shown changes in the activity of alkaline phosphatase and particularly intensive desquamation of the intestinal epithelium. This short communication does not reveal the state of other enzymes localized in the zone of the brush border.

In 1965–66 a series of investigations of the Laboratory of Physiological Chemistry of the University of Lund were published in which several intestinal dipeptidases, their distribution along the small intestine in man and certain animals, ontogenetic development and physiological localization were characterized in detail.

Josefsson and Sjöstrom (1966) showed that the dipeptidases, as we have previously observed, are bound to the outer surface of the cell membranes. It is extremely important that this binding is considerably less stable than, for instance, that for disaccharidases. Comparing these data with others cited in this book leads to the assumption that there are several types of binding which ensure a stable or weak fixation of enzymes which perform the membrane digestion, with the lipoprotein complexes of the membrane.

Kazaryan and Kulminskaya (1966) investigated the correlation of the migration of the epithelial cells from the crypts of the villi, synthesis of RNA (by incorporation of uridine-H^3) and the synthesis of protein (by incorporation of lysine-H^3). They found that the synthesis of RNA occurs only in the cells of the crypts and is practically absent in the villus. At the same time protein synthesis occurred in the entire epithelium, although in the crypts it was about twice that in the villus. Thus, the synthesis of protein taking place in the villus is determined by the presynthesized m-RNA which was confirmed by special experiments employing actinomycin which almost completely inhibited the synthesis of RNA but had no effect on the incorporation of lysine-H^3 into the protein.

Employing a very large resolution of the electron microscope, Ito (1965) made an important advance in the characterization of the ultrastructure of the microvilli. He showed that the surface of the microvilli is not smooth but has a considerable number of filaments. The author suggests that these ultrastructures may have a direct relationship to membrane digestion.

By electron microscopy of the isolated brush border of the epithelial cells of the small intestine, Johnston (1967) observed numerous knobs on the outer and laminal surface of the membrane of the microvilli with a diameter of 60 Å. There is reason to believe that these knobs are enzymes or more complex structures which include enzymes, particularly invertase and maltase. Treatment of the membranes with papain removed the knobs and along with them the corresponding enzymatic activities.

Recently a series of reports appeared which permit the characterization of the role of the intestinal surface in hydrolysis of lipids better than heretofore Hofmann (1966); Isselbacher (1966); Masarei and Simmonds (1966); Rubin (1966). For instance, during a thorough investigation (Masarei and Simmonds, 1966) of the absorption of fat and fatty acids in the small intestine of rats with bile and pancreatic fistulae it was found that the emulsified olive oil was not absorbed in the absence of bile and pancreatic juice, although the addition of taurocholate and bicarbonate restored the absorption to normal levels in spite of the absence of the enzymes of the pancreas (oleic acid was absorbed under all conditions). It is of interest that in animals with bile–pancreatic duct fistulae the contents of the intestine had only slight lipolytic activity. The authors draw attention to the fact that the origin of the lipolytic activity in rats with bile–pancreatic duct fistulae is not clear.

Chernyaskhovskii and Ugolev (1966) made an attempt to analyze the localization of the final stages of hydrolysis of tributyrine in the small intestine of white rats. It was established that in adult animals the initial stages of hydrolysis of tributyrine are accomplished mainly in the cavity of the small intestine and partially on its surface (65–75% and 25–35%, respectively). The final stage of hydrolysis (as determined by the increase in free glycerol) is accomplished by the enzymes bound to the structures of the intestinal cells [possibly identical with those described by Pope et al. (1966)] and is practically absent during the incubation with the intestinal contents. The rate of complete hydrolysis of tributyrine by homogenates and cells prepared from these homogenates under identical conditions (prepared by special techniques) was also investigated. We assumed that such an approach would permit us to determine the correlation of the lipase activity on the cell surface and inside the cells. It was found that during short incubations the rate of increase in both cases was about the same, while during more prolonged incubations the hydrolysis by the intact cells was about 70–80% of the activity of homogenates. Analogous data were obtained during the investigation of the distribution of free glycerol which results from complete hydrolysis of tributyrine. During short incubations an increase in free glycerol inside the cells was not observed, while in the surrounding medium free glycerol was present in considerable amounts. With continuation of incubation the amount of the intracellular glycerol increases, becoming even higher than in the incubation medium. The data obtained suggest that in adult rats the initial stages of the hydrolysis of triglycerides with short chains takes place in the intestinal cavity, and its final stages take place on the surface of the small intestine.

In the section "Topography of Membrane Digestion" (Chapter IV) we expressed the hypothesis that a competition may exist between the various components of complex food products for the surface of the intestinal epithelium or its definite sector. Recently we performed direct experiments (Ugolev et al., 1967) to resolve this question. The hydrolysis of glycyl-leucine, sucrose, and tributyrine, as well as combinations of these compounds, was studied. It was established that the hydrolysis of the dipeptide was sharply inhibited by tributyrine and was somewhat increased by sucrose; on the other hand, the hydrolysis of tributyrine was increased in the presence of glycyl-L-leucine. The cleavage of sucrose was not affected by the presence of the dipeptide and, under certain conditions, of tributyrine. Thus, the interaction of various food substrates on each other during membrane hydrolysis is a reality which is of significance for the understanding of the normal or altered activity of the small intestine, the origin of the proximal–distal gradient, etc. At the same time this interaction cannot be attributed only to competition, as we assumed earlier.

No information was available up to the present time regarding the distribution of enzymes between the cavity of the small intestine and its surface during digestion in dogs. Such data are the more important since dogs remain the most convenient objects of gastroenterological experiments.

Table 13 describes to a certain extent the topography of the digestive processes in dogs three hours after feeding meat.

De Laey (1966, 1967) published six communications which elucidate various aspects of reversible adsorption of pancreatic amylase on the surface of the small intestine of white rats, membrane hydrolysis of starch, etc. Of significance is the nonidentity of the effects of the same substance, soluble amylase (sA) and that adsorbed on the surface (aA). Thus, oleic and stearic acids increase the activity of aA but not of sA. Gluten induces a definite inhibition of hydrolysis of starch in the cavity of the intestine as well as on its surface, while casein and milk decrease the intensity of only the membrane digestion. De Laey, as well as ourselves, found an increase in the specific activity of α-amylase during its adsorption and a change in the stereochemical state of the adsorbed enzyme which is reflected, in part, in higher stability toward effects of pH and temperature. The energy of activation of the adsorbed amylase is lower than of the dissolved enzyme. The data obtained by the author during the investigation of the effects of various cationic detergents (for example, cetrimide or tetraethylammonium bromide) and anionic detergents (lauryl sulfate or aerosol OT) are essential for the understanding of the mechanisms of exchange of the enzymes between the cavity and the surface of the small intestine. These data demonstrate the possible role of the phospholipids of the membrane in the processes of adsorption of the enzymes. The important significance of the basic groups of the membrane is indicated by the competition between lauryl sulfate and amylase during adsorption. The activating effect of cetrimide (in concentrations smaller than the critical micellar concentration) indicates

TABLE 13

Distribution of Certain Digestive Enzymes between the Contents of the Small Intestine and Its Mucosa in Dogs Three Hours after Consumption of Cooked Meat*

No. of animal	Object of study	Amylase	Invertase	Glycyl-leucine-dipeptidase	Tributyrinase	Monoglyceride lipase
1	Contents	1071.1(52.5)	920.5(6.4)	153.9(10.8)	645.86(73.2)	1.866(25.4)
	Mucosa	967.9(47.5)	13485.7(93.6)	1272.1(89.2)	234.7(26.8)	5.48(74.6)
		2039.9(100)	14406.2(100)	1426.0(100)	880.56(100)	7.346(100)
2	Contents	–	594.3(5.3)	4.8(0.3)	467.3(48)	0.874(10.3)
	Mucosa	–	10693.5(94.7)	1805.2(99.7)	508.3(52)	7.567(89.7)
		–	11287.8(100)	1810(100)	975.6(100)	8.441(100)
3	Contents	8994.0(69.6)	786.5(4.1)	15.8(1.04)	–	–
	Mucosa	3924.5(30.4)	18158.6(95.9)	1513.5(98.06)	–	–
		12918.5(100)	18945.1(100)	1529.3(100)	–	–

*The activities were determined in absolute units – in milligrams per minute and in percent (in parentheses). Absolute activity was assayed: for amylase by amount of hydrolyzed starch; for invertase by formation of glucose; for glycyl-leucine-dipeptidase by formation of glycine; and for tributyrinase and monoglyceride lipase by amount of cleaved tributyrine (Iezuitova et al., 1967).

the significance of the fluctuations of local surface tension for the changes in the activity of adsorbed enzymes.

Unfortunately, in our book insufficient attention was given to the superb work of F. Moog and co-workers devoted to the regulation of the phosphatase activity of the microvilli (see review: 1965).

Masevich et al. (1967) attempted to develop a method which is convenient for the determination of desorption of amylase from the surface of the small intestine under clinical conditions. In the final analysis this method amounted to sequential incubation of pieces of small intestine obtained from biopsy in a cooled Ringer solution, after which the residual activity of the homogenate was determined. It turned out that the correlation of amylolytic activity in various fractions was substantially altered depending on the character of the disease. By simplification to a certain extent the authors consider the first fraction as a mixture of the soluble and easily desorbed amylase, the second fraction as easily desorbed amylase, the 3rd and 4th as the average-desorbed amylase, and the 5th as the undesorbable or difficult to desorb amylase.

It was found that in a deficiency of the pancreatic gland there could be a sharp rise in the content of amylase in the 3rd to the 5th fractions. On the contrary, in various enteritis conditions (in many cases accompanied by histological alterations of the structure of the villi) there is often observed a sharp decrease in the content of amylase in the 3rd to the 4th and particularly in the 5th fractions. It is possible that the observed phenomena are not only of diagnostic interest, but could be considered as a reflection of peculiar compensatory reactions.

In essence a similar method was employed by Ustinkova and Ugolev for the study of certain features of desorption of amylase in experiments on rats. Instead of small pieces of intestine, however, we used inverted slices of the small intestine of about 5 cm in length. More importantly, all the amylase initially present was first desorbed by the incubation of the inverted slices in a cooled Ringer solution for about one hour. It was found in these experiments that the total content of amylase adsorbed at 38°C is higher than that adsorbed at 0°. It is of particular interest that the content of the average-desorbable and difficult-to-desorb amylase rises sharply. If only the physical adsorption of the enzyme were involved here one could expect the opposite results. It could thus be assumed that amylase, at least in part, is retained on the surface of the small intestine not by physical but by chemical adsorption. Possibly, this explains the considerable influence of the functional state of the small intestine on the amount of bound enzymes and the stability of their binding to the surface of the cells. It is not excluded that the processes of adsorption and desorption are determined not only by the presence of groupings which enter into chemical bond with the enzyme but also by the geometric properties of various adsorption centers. Certain differences in the kinetics of adsorption by the slices of the thin intestine of rats of rat and dog amylase indirectly suggest such a possibility.

Bibliography

Adam, N. K., 1947. Physics and chemistry of surfaces. Gostekhizdat, Moscow–Leningrad.

Agar, W.T., Hird, F.J.R., and Sidhu, G.S., 1953. The active absorption of amino acids by the intestine. J. Physiol., Vol. 121, No. 2, pp. 255–263.

Agar, W.T., Hird, F.J.R., and Sidhu, G.S., 1954. The uptake of amino acids by the intestine. Biochim. Biophys. Acta, Vol. 14, No. 1, pp. 80–84.

Alvarez, A., and Sas, J., 1961. β-Galactosidase changes in the developing intestinal tract of the rat. Nature, Vol. 190, No. 4778, pp. 826–827.

Arshavskii, I.A., 1957. Specific peculiarities of the physiology of digestion in early age periods. Pediatriya, No. 7, pp. 24–26.

Ashworth, C.T., and Chears, W.C., 1962. Follow-up of intestinal biopsy in nontropical sprue after gluten-free diet and remission. Feder. Proc., Vol. 21, No. 6, pp. 880–890.

Ashworth, C.T., Luibel, F.J., and Stewart, S.C., 1963. The fine structural localization of adenosine triphosphatase in the small intestine, kidney, and liver of the rat. J. Cell. Biol., Vol. 17, No. 1, pp. 1–18.

Auricchio, S., Rubino, A., Landolt, M., Semenza, G., and Prader, A., 1963. Isolated intestinal lactase deficiency in the adult. Lancet, Vol. 2, No. 7303, pp. 324–326.

Auricchio, S., Rubino, A., Prader, A., Rey, J., Jos, J., Frézal, J., and Davidson, M., 1965. Intestinal glycosidase activities in congenital malabsorption of disaccharides. J. Pediat., Vol. 66, No. 3, pp. 555–564.

Auricchio, S., Semenza, G., and Rubino, A., 1965. Multiplicity of human intestinal disaccharidases. II. Characterization of the individual maltases. Biochim. Biophys. Acta, Vol. 96, No. 3, pp. 498–507.

Babkin, B.P., 1927. External secretion of digestive glands. Gosizdat, Moscow–Leningrad.

Babkin, B.P., 1960. Secretory mechanism of digestive glands. Medgiz, Leningrad.

Bailey, C.B., Kitts, W.D., and Wood, A.J., 1956. The development of the digestive enzyme system of the pig during its preventing phase of growth. V. Intestinal lactase, sucrase, and saltase. Can. J. Agr. Sci., Vol. 36, No. 1, p. 51.

Baker, S.J., and Hughes, A., 1960. Multiple-retrieving small-intestinal biopsy tube. Lancet, Vol. 2, No. 7152, pp. 686–687.

Baker, S.J., Mathan, V.I., and Cherian, V., 1963. The nature of the villi in the small intestine of the rat. Lancet, Vol. 1, No. 7286, p. 860.

Barcroft, D.K., 1937. Basic character of the architecture of physiological functions. Biomedgiz, Moscow–Leningrad.

Bardare, M., Buitoni, P., and Careddu, P., 1965. Modificazioni della curva glicemica dopo carico con lattosio nei bambini normali e distrifici. Minerva Pediat., Vol. 17, No. 14, pp. 832–835.

Bar-Eli, A., and Katchalski, E., 1960. A water-insoluble trypsin derivative and its use as a trypsin column. Nature, Vol. 188, No. 4753, pp. 856–857.

Bar-Eli, A., and Katchalski, E., 1963. Preparation and properties of water-insoluble derivatives of trypsin. J. Biol. Chem., Vol. 238, No. 5, pp. 1690–1698.

Barrington, E.J.W., 1962. Digestive enzymes. In: Advances in Comparative Physiology and Biochemistry, Vol. 1. New York–London, pp. 1–65.

Barrnett, R.J., 1959. The demonstration with the electron microscope of the end-products of histochemical reactions in relation to the fine structure of cells. Exptl. Cell Research, Suppl. 7 (The cytochemistry of enzymes and antigens), pp. 65–89.

Becker, D.E., Ullrey, D.E., Terrill, S.W., and Notzold, R.A., 1954. Failure of the newborn pig to utilize dietary sucrose. Science, Vol. 120, No. 3113, pp. 345–346.

Beeken, W.L., and Norman, M., 1964. Degradation of albumin by isolated segments of the gastrointestinal tract in rats. Proc. Soc. Exptl. Biol. Med., Vol. 117, No. 1, pp. 24–27.

Berman, Sh. A., 1964. Materials on contract digestion in carp of two different ages. XI. Scientific Conference for the Study of Internal Water Basins. Petrozavodsk, pp. 7–8.

Bernal, G., 1959. Problem of stages in biopoiesis. In: Origin of Life on Earth (Materials of International Symposium, 19–24 August, 1957). Izd. AN SSSR, Moscow, pp. 42–58.

Bernhard, K., Wagner, H., and Ritzel, G., 1952. Versuche zur quantitativen Erfassung der bei der Resorption von Neutralfett eintretenden Spaltung. Helv. Chim. Acta, Vol. 35, No. 4, pp. 1404–1411.

Bertalanffy, F.D., and Nagy, K., 1958. The mitotic activity and renewal rate of human duodenal epithelium. Anat. Record, Vol. 130, No. 2, pp. 271–272.

Best, J.B., 1955. The inference of intracellular enzymatic properties from kinetic data obtained on living cells. II. A study of hexokinase and invertase as cellular components of baker's yeast. J. Cell. Comp. Physiol., Vol. 46, No. 1, pp. 29–52.

Biempica, L., Vilar, O., and Gorodisch, S., 1961. Histochemical study of the enzymatic activity of the normal small intestine epithelium obtained by aspiration biopsy. Anat. Record, Vol. 139, No. 2, p. 294.

Bierry, H., 1912. Saccharose spaltende Fermente. Biochem. Z., Vol. 44, pp. 415–425.

Blair, D.G.R., and Tuba, J., 1963. Rat intestinal sucrase. I. Intestinal distribution and reaction kinetics. Can. J. Biochem. Physiol., Vol. 41, No. 4, pp. 905–916.

Blair, D.G.R., Yakimets, W., and Tuba, J., 1963. Rat intestinal sucrase. II. The effects of rat age and sex and of diet on sucrase activity. Can. J. Biochem. Physiol., Vol. 41, No. 4, pp. 917–929.

Bochkov, N.P., 1958. Morphophysiological changes in dog small intestine after partial resection. Bull. Exptl. Biol. Med., Vol. 46, No. 10, pp. 104–108.

Bockus, H.L., 1964. Gastroenterology, Vol. 2. Philadelphia–London.

Bogatch, P.G., 1961. Mechanisms of nervous regulation of peristalsis of small intestine. Kiev State University, Kiev.

Bolt, R.J., 1964. Methods of small bowel biopsy. J. Am. Med. Assoc., Vol. 188, No. 1, pp. 40–41.

Bolt, R.J., Pollard, H.M., and McCool, S., 1960. Staining of enzymes in mucosa of the small bowel, using a peroral biopsy tube. Am. J. Clin. Pathol., Vol. 34, No. 1, pp. 43–49.

Bolton, C., and Wright, G.P., 1937. The absorption of amino acids and their distribution in the body fluids. J. Physiol., Vol. 89, No. 3, pp. 269–286.

Bonneville, M.A., 1963. Fine structural changes in the intestinal epithelium of the bullfrog during metamorphosis. J. Cell. Biol., Vol. 18, No. 3, pp. 579–597.

Booij, H.L., 1962. Colloid chemistry of living membranes. Conf. on Permeability. Wageningen, pp. 5–35.

Borgström, B., 1952. On the action of pancreatic lipase on triglycerides in vivo and in vitro. Acta Physiol. Scand., Vol. 25, No. 4, pp. 328–347.

Borgström, B., 1954. Effect of tauro-cholic acid on the pH/activity curve of rat pancreatic lipase. Biochim. Biophys. Acta, Vol. 13, No. 1, pp. 149–150.

Borgström, B., and Dahlqvist, A., 1958. Cellular localization, solubilization and separation of intestinal glycosidases. Acta Chem. Scand., Vol. 12, No. 10, pp. 1997–2006.

Borgström, B., Dahlqvist, A., Lundh, G., and Sjövall, J., 1957. Studies of intestinal digestion and absorption in the human. J. Clin. Invest., Vol. 36, No. 10, pp. 1521–1536.

Borgström, B., Lundh, G., and Hofmann, A., 1963. The site of absorption of conjugated bile salts in man. Gastroenterology, Vol. 45, No. 2, pp. 229–238.

Bourne, G., 1943. Distribution of alkaline phosphatase in various tissues. Quart. J. Exptl. Physiol., Vol. 32, No. 1, pp. 1–19.

Brachet, J., and Mirsky, A.E. (eds.), 1961. The Cell. Biochemistry, Physiology, Morphology. Vol. 2. Cells and Their Component Parts. New York–London.

Bradfield, J.R.G., 1950. The localization of enzymes in cells. Biol. Rev., Vol. 25, No. 1, pp. 113–157.

Brambell, F.W.R., 1958. The passive immunity of the young mammal. Biol. Rev., Vol. 33, No. 4, pp. 488–531.

Brandborg, L.L., Rubin, G.E., and Quinton, W.E., 1959. A multipurpose instrument for suction of the esophagus, stomach, small bowel, and colon. Gastroenterology, Vol. 37, No. 1, pp. 1–16.

Brandes, D., Zetterqvist, H., and Sheldon, H., 1956. Histochemical techniques for electron microscopy. Alkaline phosphatase. Nature, Vol. 177, No. 4504, pp. 382–383.

Braunshtein, A.E., 1963. "Active centers" and the nature of catalytic activities of enzymes. D.I. Mendeleev Journal of All-Union Chemical Society, Vol. 8, No. 1, pp. 81–94.

Braunshtein, A.E., Karpeiskii, M. Ya., and Khomutov, R.M., 1964. Recent views on the mechanism of the catalytic effects of enzymes. In: Enzymes (The fundamentals of molecular biology). Nauka, Moscow, pp. 237–268.

Brenner, D.M., 1962. On the question of the state of the gastrointestinal tract in chronic saturnism. Radiologiya, Vol. 3, No. 9, pp. 323–328.

Brodskii, R.A., 1962. Changes in the content of RNA and in the activity of alkaline phosphatase in the mucosa of the small intestine of white rats and rabbits in early postnatal period. Arch. Anat., Gistol. i Embriol., Vol. 72, No. 3, pp. 92–102.

Brodskii, R.A., 1964. Histochemical methods of study of resorption processes. Symposium: Physiology and Pathology of Absorption in Gastrointestinal Tract. Odessa, pp. 150–160.

Brown, A.L., 1962. Microvilli of the human jejunal epithelial cell. J. Cell. Biol., Vol. 12, No. 3, pp. 623–627.

Brown, R., 1955. Enzymes of the surface of protoplast and absorption of sugar. In: Modern Problems of Cytology. IIL, Moscow, pp. 213–225.

Buddenbrock, W., 1956. Vergleichende Physiologie, Vol. 3. Basel–Stuttgart.

Buensod, M., Favarger, P., and Collet, R.A., 1953. Le sort du glycérol libre et du glycérophosphate pendant la résorption des graisses. Helv. Physiol. Pharmacol. Acta, Vol. 11, No. 1, pp. 45–48.

Bülbring, E., 1961. Motility of the intestine. Proc. Roy. Soc. Med., Vol. 54, No. 9, pp. 773–775.

Burstone, M.S., and Weisburger, E.K., 1961. Histochemical demonstration of aminopeptidase by a new oxidation technique. J. Histochem. Cytochem., Vol. 9, No. 6, pp. 712–713.

Burstone, M.S., 1965. Enzyme Histochemistry. Mir, Moscow. [English edition: 1962, Academic Press, New York.]

Butterworth, C.E., and Perez-Santiago, E., 1958. Jejunal biopsies in sprue. Ann. Internal Med., Vol. 48, No. 1, pp. 8–29.

Bykov, K.M., 1940. Lectures on the Physiology of Digestion. Voyenno-Moskaya Meditsinskaya Akademiya, Leningrad.

Bykov, K.M., and Ryccle, A.V., 1935. The role of mucus in the process of digestion. In: Neurohumoral Regulations in the Activity of the Digestive Apparatus in Man. Ed. K.M. Bykov, Vol. 2. VIEM, Moscow, pp. 153–162.

Cajori, F.A., 1933. The enzyme activity of dog's intestinal juice and its relation to intestinal digestion. Am. J. Physiol., Vol. 104, No. 3, pp. 659–668.

Catrecasas, P., Lockwood, D.H., and Caldwell, J.R., 1965. Lactase deficiency in the adult, A common occurrence. Lancet, Vol. 1, No. 7375, pp. 14–22.

Cebra, J.J., Givol, D., Silman, H.I., and Katchalski, E., 1961. A two-stage cleavage of rabbit γ-globulin by a water-insoluble papain preparation followed by cysteine. J. Biol. Chem., Vol. 236, No. 6, pp. 1720–1725.

Cevini, G., Ciovannini, M., and Careddu, P., 1962. Alterazioni della digestione e dell'assorbimento intestinale del glucidi nei disturbi acuti e cronici della nutrizione del lattante. Minerva Pediat., Vol. 14, Nos. 31–32, pp. 831–835.

Chain, E.B., Mansford, K.R.L., and Pocchiari, F., 1960. The absorption of sucrose, maltose, and higher oligosaccharides from the isolated rat small intestine. J. Physiol., Vol. 154, No. 1, pp. 39–51.

Chase, W.H., 1963. The demonstration of alkaline phosphatase activity in frozen-dried mouse gut in the electron microscope. J. Histochem. Cytochem., Vol. 11, No. 1, pp. 96–101.

Chernyaskhovskii, M.Yu., and Ugolev, A.M., 1966. Reports of the XVI Scientific Session of the Institute of Nutrition of the USSR Academy of Medical Sciences, Vol. 1, pp. 146–147.

Clark, S.L., 1961a. Alkaline phosphatase of the small intestine studied with electron microscope in suckling and adult mice. Anat. Record, Vol. 139, No. 2, p. 216.

Clark, S.L., 1961b. The localization of alkaline phosphatase in tissues of mice, using the electron microscope. Am. J. Anat., Vol. 109, No. 1, pp. 57–62.

Clarkson, E.M., and Maizels, M., 1952. Distribution of phosphatases in human erythrocytes. J. Physiol., Vol. 116, No. 1, pp. 112–128.

Clément, G., 1964. La digestion et l'absorption des graisses. J. Physiol., Vol. 56, No. 1, pp. 111–192.

Clément, G., Clément, J., and Bezard, J., 1962. Action of human pancreatic lipase on synthetic mixed symmetrical triglycerides of long-chain acids and butyric acid. Biochem. Biophys. Res. Comm., Vol. 8, No. 3, pp. 238–242.

Connel, A.M., 1961. The motility of the small intestine. Postgrad. Med. J., Vol. 37, No. 434, pp. 703–716.

Conrad, R.A., 1956. Some effects of ionizing radiation on the physiology of the gastrointestinal tract: a review. Rad. Res., Vol. 5, No. 2, pp. 167–188.

Crane, R.K., 1960. Intestinal absorption of sugars. Physiol. Reviews, Vol. 40, No. 4, pp. 789–825.

Crane, R.K., 1962. Hypothesis for mechanism of intestinal active transport of sugars. Feder. Proc., Vol. 21, No. 6, pp. 891–895.

Crane, R.K., 1965. Na^+-dependent transport in the intestine and other animal tissues. Feder. Proc., Vol. 24, No. 5, pp. 1000–1006.

Crane, R.K., 1966. Enzymes and malabsorption: a concept of brush border membrane disease. Gastroenterology, Vol. 50, No. 2, pp. 254–262.

Crane, C.W., and Neuberger, A., 1960. The digestion and absorption of protein by normal man. Biochem. J., Vol. 74, No. 2, pp. 313–323.

Crimm, P.D., and Strayer, J.W., 1936. Phosphatase content of blood serum and tissues in the rat following administration of vitamins D and A. J. Biol. Chem., Vol. 112, No. 2, pp. 511–515.

Crosby, W.H., 1963. Small intestinal studies. Methods for obtaining intraluminal contents and intestinal mucosa. Am. J. Digest. Diseases, Vol. 8, No. 1, pp. 2–11.

Crosby, W.H., and Kugler, H.W., 1957. Intraluminal biopsy of the small intestine. The intestinal biopsy capsule. Am. J. Digest. Diseases, New Series, Vol. 2, No. 5, pp. 236–241.

Csáky, T.Z., 1964. Intestinal absorption. Changing concept and ideas. Gastroenterology, Vol. 47, No. 2, pp. 201–205.

Csáky, T.Z., and Fernald, G.W., 1961. Localization of the "sugar pump" in the intestinal epithelium. Nature, Vol. 191, No. 4789, pp. 709–710.

Curran, P.F., Webster, E.W., and Hovsepian, J.A., 1960. The effect of x-irradiation on sodium and water transport in rat ileum. Rad. Res., Vol. 13, No. 2, pp. 369–380.

Dahlqvist, A., 1960. Hog intestinal α-glucosidases. Solubilization, separation and characterization. [Lund]

Dahlqvist, A., 1962. The intestinal disaccharidases and disaccharide intolerance. Gastroenterology, Vol. 43, No. 6, pp. 694–696.

Dahlqvist, A., and Borgström, B., 1961. Digestion and absorption of disaccharides in man. Biochem. J., Vol. 81, No. 2, pp. 411–418.

Dahlqvist, A., and Brun, A., 1962. A method for the histochemical demonstration of disaccharidase activities: application to invertase and trehalase in some animal tissues. J. Histochem. Cytochem., Vol. 10, No. 3, pp. 294–302.

Dahlqvist, A., and Thomson, D.L., 1963a. The digestion and absorption of sucrose by the intact rat. J. Physiol., Vol. 167, No. 2, pp. 193–209.

Dahlqvist, A., and Thomson, D.L., 1963b. The digestion and absorption of maltose and trehalose by the intact rat. Acta Physiol. Scand., Vol. 59, No. 1–2, pp. 111–125.

Dahlqvist, A., Hammond, J.B., Crane, R.K., Dunphy, J.V., and Littman, A., 1963. Intestinal lactase deficiency and lactose intolerance in adults. Gastroenterology, Vol. 45, No. 4, pp. 488–491.

Dalton, A.J., Kahler, H., and Lloyd, B.J., 1951. The structure of the free surface of a series of epithelial cell types in the mouse as revealed by the electron microscope. Anat. Record, Vol. 111, No. 1, pp. 67–72.

Dalton, A.J., Kahler, H., Striebich, M.J., and Lloyd, B., 1950. Finer structure of hepatic, intestinal, and renal cells of the mouse as revealed by the electron microscope. J. Nat. Cancer Inst., Vol. 11, No. 2, pp. 439–443.

Danielli, J.F., 1952. Structural factors in cell permeability and secretion. In: Structural Aspects of Cell Physiology (Symposia of the Soc. for Exptl. Biol., No. VI). Cambridge, pp. 1–15.

Danielli, J.F., 1954a. Morphological and molecular aspects of active transport. In: Active Transport and Secretion (Symposia of the Soc. for Exptl. Biol., No. VIII). Cambridge, pp. 502–516.

Danielli, J.F., 1954b. The present position in the field of facilitated diffusion and selective active transport. In: Recent Developments in Cell Physiology (Proc. VII Symp. Colston Res. Soc. Univ. of Bristol). London, pp. 1–14.

Deane, H.W., and Dempsey, E.W., 1945. The localization of phosphatases in the Golgi region of intestinal and other epithelial cells. Anat. Record, Vol. 93, p. 401.

De Boer, J.H., 1959. Phenomena of adsorption. In: Catalysis. Certain Aspects of Theory and Technology of Organic Reactions. IIL, Moscow, pp. 18–176.

De Boer, J.H., 1962. The Dynamic Nature of Adsorption. IIL, Moscow. [English edition: 1953, Oxford University Press.]

Deborin, G.A., and Gorbacheva, L.B., 1953. Complexes of proteins with lipoids and their properties. Biokhimiya, Vol. 18, No. 5, pp. 618–625.

Deborin, G.A., Bystrova, M.I., and Ivanova, V.P., 1959. Changes in the course of proteolysis of serum albumin by trypsin during formation of the complexes of the enzyme or substrate with estradiol. Dokl. Akad. Nauk SSSR, Vol. 124, No. 3, pp. 685–687.

De Duve, C., 1959. Subcellular Particles. New York.

De Groot, A.P., and Hoogendoorn, P., 1957. The detrimental effect of lactose. II. Quantitative lactase determinations in various mammals. Neth. Milk Dairy J., Vol. 2, pp. 290–303.

De Laey, P., 1963. Personal communication.

De Laey, P., 1966. Mahrung, Vol. 10, No. 8, pp. 641, 649, 655.

De Laey, P., 1967. Mahrung, Vol. 11, No. 1, pp. 1, 9, 17.

De Laey, P., and Iezuitova, N.N., 1962. On the mechanism of the enzymatic cleavage of starch in the intestine. Dokl. Akad. Nauk SSSR, Vol. 146, No. 3, pp. 731–733.

De La Fuente, G., and Sols, A., 1962. Transport of sugars in yeasts. II. Mechanisms of utilization of disaccharides and related glycosides. Biochem. Biophys. Acta, Vol. 56, No. 1, pp. 49–62.

Demis, D.J., Rothstein, A., and Meier, R., 1954. The relationship of the cell surface to metabolism. X. The location and function of invertase in the yeast cell. Arch. Biochem. Biophys., Vol. 48, No. 1, pp. 55–62.

Dempsey, E.W., and Deane, H.W., 1946. The cytological localization, substrate specificity, and pH optima of phosphatases in the duodenum of the mouse. J. Cell. Compar. Physiol., Vol. 27, No. 3, pp. 159–171.

Denton, A.E., and Elvehjem, C.A., 1954. Availability of amino acids in vivo. J. Biol. Chem., Vol. 206, No. 1, pp. 449–454.

Denton, A.E., Gershoff, S.N., and Elvehjem, C.A., 1953. A new method for cannulating the portal vein of dogs. J. Biol. Chem., Vol. 204, No. 2, pp. 731–735.

Desnuelle, P., and Constantin, M.J., 1952. Formation de glycérides partiels pendant la lypolyse des triglycérides dans l'intestin. Biochim. Biophys. Acta, Vol. 9, No. 5, pp. 531–537.

Desnuelle, P., Naudet, M., and Constantin, M.J., 1951. Sur un nouveau type de lipolyse in vitro, une lipolyse génératrice de glycérol. Biochim. Biophys. Acta, Vol. 7, No. 3, pp. 251–256.

Desnuelle, P., Naudet, M., and Rouzier, J., 1947. Sur la formation de glycérides partiels au cours de l'hydrolyse des triglycérides par la lipase pancréatique. Compt. Rend. Soc. Biol., Vol. 141, No. 23–24, pp. 1242–1244.

Desnuelle, P., Naudet, M., and Rouzier, J., 1948. Etude quantitative de la formation de glycérides partiels au cours de l'hydrolyse fermentaire de triglycérides. Biochim. Biophys. Acta, Vol. 2, No. 6, pp. 561–574.

Desnuelle, P., Sarda, L., and Ailhaud, G., 1960. Inhibition de la lipase pancréatique par le diéthyl-p-nitrophényl phosphate en emulsion. Biochim. Biophys. Acta, Vol. 37, No. 3, pp. 570–571.

Detrick, L.E., Latta, H., Upham, H.C., and McCandless, R., 1963. Electron-microscopic changes across irradiated rat intestinal villi. Rad. Res., Vol. 19, No. 3, pp. 447–461.

Detrick, L.E., Upham, H.C., Dunlap, A.K., and Haley, T.J., 1961. Thiamine transport across the irradiated isolated rat intestine. Rad. Res., Vol. 15, No. 4, pp. 520–526.

Dickson, H.M., 1955. Effect of x-irradiation on glucose absorption. Am. J. Physiol., Vol. 182, No. 3, pp. 477–478.

Dinischiotu, Ç.T., Rădulescu, I.C., Crăciun, O., and Prida, N., 1960. Modificarile radiologice ale tractuli gastrointestinal in saturnism. Viata Med., No. 3, pp. 213–220.

Diskina, B.S., 1960. Effect of the configuration of DNA molecule on the biological activity of proteins bound to DNA. Biokhimiya, Vol. 25, No. 1, pp. 43–54.

Doell, R.G., and Kretchmer, N., 1962. Studies of small intestine during development. I. Distribution and activity of β-galactosidase. Biochim. Biophys. Acta, Vol. 62, No. 2, pp. 353–362.

Doell, R.G., and Kretchmer, N., 1963. Invertase in the intestine of the developing rat. Feder. Proc., Vol. 22, No. 2, Part I, p. 495.

Doell, R.G., and Kretchmer, N., 1964. Intestinal invertase. Precocious development of activity after injection of hydrocortisone. Science, Vol. 143, No. 3601, pp. 42–44.

Doell, R.G., Rosen, G., and Kretchmer, N., 1965. Immunochemical studies of intestinal disaccharidases during normal and precocious development. Proc. Nat. Acad. Sci. USA, Vol. 54, pp. 1286–1273.

Dombrovskaya, M.P., Sankina, N.V., and Khomitskaya, T.A., 1965. The nutrition and vitamin therapy of children with exema. Abstracts, III All-Union Conference of Pediatricians. Volgograd, pp. 48–49.

Doniach, I., and Shiner, M., 1957. Duodenal and jejunal biopsies. II. Histology. Gastroenterology, Vol. 33, No. 1, pp. 71–86.

Douglas, G.J., Reinauer, A.J., Brooks, W.C., and Pratt, J.H., 1953. The effect on digestion and absorption of excluding the pancreatic juice from the intestine. Gastroenterology, Vol. 23, No. 3, pp. 452–459.

Dreisbach, L., and Nasset, E.S., 1954. Absorption of carbohydrate and protein as affected by feeding cornstarch, banana, or glucose. J. Nutr., Vol. 53, No. 4, pp. 523–532.

Dumesh, M., 1965. Capsule for biopsy of the mucosa of the gastrointestinal tract. Izv. Akad. Nauk Latv.SSR, No. 11, pp. 107–110.

Dunphy, J.V., Littman, A., Hammond, J.B., Forstner, G., Dahlqvist, A., and Crane, R.K., 1965. Intestinal lactase deficit in adults. Gastroenterology, Vol. 49, No. 1, pp. 12–21.

Emmel, V.M., 1946. The intracellular distribution of alkaline phosphatase activity following various methods of histologic fixation. Anat. Record, Vol. 95, No. 2, pp. 159–173.

Entressangles, B., Pasero, L., Savary, P., Sarda, L., and Desnuelle, P., 1961. Influence de la nature des chaînes sur la vitesse de leur hydrolyse par la lipase pancréatique. Bull. Soc. Chim. Biol., Vol. 43, No. 4, pp. 581–591.

Faitelberg, P.O., 1960. Absorption in Digestive Apparatus. Medgiz, Moscow.

Farrar, J.T., Small, M.D., Bullard, D., and Ingelfinger, F.J., 1956. Effect of total-body irradiation on absorption of sugars from the small intestine. Am. J. Physiol., Vol. 186, No. 3, pp. 549–553.

Favarger, P., and Collet, R.A., 1950. Estimation du degré d'hydrolyse des graisses pendant la résorption intestinale. Helv. Physiol. Pharmacol. Acta, Vol. 8, No. 2, pp. C15–C16.

Favarger, P., Collet, R.A., and Cherbuliez, E., 1951. Etude de la résorption intestinale des graisses à l'aide de deutéro-glycérol et d'acides gras marqués. Helv. Chim. Acta, Vol. 34, No. 5, pp. 1641–1654.

Filippovitch, S.I., 1962. On Adaptive Processes in Alterations of Activity of Digestive System. Medgiz, Moscow.

Filippovitch, S.I., Amirov, N. Sh., Volkova, T.V., and Martsevitch, M.S., 1960. Additional data on the physiological analysis of compensatory processes in injuries of digestive system. The Works of Scientific Conference on the Problems of Physiology and Pathology of Digestion, Dedicated to the Memory of Academician K.M. Bykov. Ivanovo, pp. 885–860.

Filippovitch, S.I., Martsevitch, M.S., Volkova, T.V., and Sabsai, B.I., 1961. On the compensation of the functions of digestion after general resection of the small intestine in dogs. Scientific Conference on the Problems of Physiology and Pathology of Digestion and Absorption. Abstracts of Reports. Odessa, pp. 261–264.

Fink, K., 1944. The effect of the thyroid on jejunal secretion in the dog. Am. J. Physiol., Vol. 141, No. 4, pp. 598–605.

Fisher, R.B., 1954. Protein Metabolism, Ch. 1. London–New York.

Fisher, R.B., and Parsons, D.S., 1953. Glucose movements across the wall of the rat small intestine. J. Physiol., Vol. 119, No. 2–3, pp. 210–223.

Fishman, W.H., Green, S., and Inglis, N.I., 1963. L-phenylalanine: an organ specific, stereospecific inhibitor of human intestinal alkaline phosphatase. Nature, Vol. 198, No. 4881, pp. 685–686.

Flick, A.L., Quinton, W.E., and Rubin, C.E., 1961. A peroral hydraulic biopsy tube for multiple sampling at any level of the gastrointestinal tract. Gastroenterology, Vol. 40, No. 1, pp. 120–126.

Floch, M.H., Caldwell, W.L., and Sheehy, T.W., 1962. A histopathologic basis for the interpretation of small bowel roentgenography in tropical sprue. Am. J. Roentgenol., Vol. 87, No. 4, pp. 709–716.

Florey, H.W., Wright, R.D., and Jennings, M.A., 1941. The secretion of the intestine. Physiol. Rev., Vol. 21, No. 1, pp. 36–69.

Florkin, M., 1947. Biochemical Evolution. IIL, Moscow. [English edition: 1949, Academic Press, New York.]

Fomina, L.S., 1951. The role of the solid portion of the intestinal secretion in the process of production of the enzymes by the intestinal mucosa. In: Problems of Nutrition. Current Problems of Nutrition of Healthy and Ill Humans. No. 1, Izd. AMN SSSR, Moscow, pp. 130–152.

Fomina, L.S., 1955a. Secretion of intestinal enzymes by the isolated and "innervated" slices of dog intestine in relation to protein content of the food. Communication I. The content of intestinal secretion. Vopr. Pitaniya, Vol. 14, No. 3, pp. 16–22.

Fomina, L.S., 1955b. Secretion of intestinal enzymes by the isolated and "innervated" slices of dog intestine in relation to protein content of food. Communication II. Secretion of intestinal enzymes and their elimination with feces. Vopr. Pitaniya, Vol. 14, No. 5, pp. 20–27.

Fomina, L.S., 1957. Secretion of intestinal enzymes. Dissertation. Moscow.

Fox, S., 1959. Chemical theory of spontaneous generation. In: Origin of Life on Earth (Proceedings of on International Symposium, 19–24 August, 1957). Izd. AN SSSR, Moscow, pp. 260–266.

Frazer, A.C., 1949. The physiology of the intestinal cell. Gastroenterologia, Vol. 75, No. 1, pp. 17–21.

Frazer, A.C. (ed.), 1963. Biochemical Problems of Lipids. Amsterdam–London–New York.

Frazer, A.C., and Sammons, H.G., 1945. The formation of mono- and diglycerides during the hydrolysis of triglyceride by pancreatic lipase. Biochem. J., Vol. 39, No. 2, pp. 122–128.

Frazer, A.C., Schneider, R., and Morg, D.B., 1963. Gluten-induced enteropathy and protein digestion. Proc. VI Internat. Congr. Nutr., Edinburg. Edinburg–London, pp. 353–357.

Fredricsson, B., and Wirsén, C., 1956. In vivo effect of colchicine on alkaline phosphatase of rat intestinal epithelium. Exptl. Cell Res., Vol. 10, No. 3, pp. 749–751.

Frolkis, A.V., 1964. Functional Interdependence of Intestine and Stomach. Meditsina, Leningrad.

Gadjieva, Z.M., 1965. Certain regularities in the division of epithelial cells in the animal small intestine. Vopr. Pitaniya, Vol. 24, No. 1, pp. 66–71.

Gallo, L.L., and Treadwell, C.R., 1963. Localization of cholesterol esterase and cholestrol in mucosal fractions of rat small intestine. Proc. Soc. Exptl. Biol. Med., Vol. 114, No. 1, pp. 69–72.

Garber, N., and Citri, N., 1966. Biochim. Biophys. Acta, Vol. 62, pp. 385–396.

Gartier, E.E., 1900. Pancreas of the embryos and the newborn of man. Dissertation. St. Petersburg.

Gavriton, R.I., and Damanskii, V. Yu., 1966. Radiobiologiya, Vol. 6, pp. 394–398.

Germain, J., 1961. Heterogeneous Catalysis. IIL, Moscow.

Ginetsinskaya, T.A., Mashanskii, V.F., and Dobrovolskii, A.A., 1966. The ultrastructure integument and method of nourishment of sporocytes (Trematodes). Dokl. Akad. Nauk SSSR, Vol. 166, No. 4, pp. 1003–1004.

Ginetsinskii, A.G., 1963. The Physiological Mechanisms of Water-Salt Equilibrium. Izd. AN SSSR, Moscow–Leningrad.

Gitzelmann, R., Davidson, E.A., and Osinchak, J., 1964. Disaccharidase of rabbit small intestine. Intracellular distribution, solubilization, purification, and specificity. Biochim. Biophys. Acta, Vol. 85, No. 1, pp. 69–81.

Goldberg, O.I., 1913. A study of the activity of the gastrointestinal tract during exclusion of the external secretion of the pancreas. Dissertation. St. Petersburg.

Goldin, L.S., 1956. Electron microscopy of the mucosa of the small intestine. Dokl. Akad. Nauk SSSR, Vol. 106, No. 4, pp. 721–724.

Gomori, G., 1939. Microtechnical demonstration of phosphatase in tissue sections. Proc. Soc. Exptl. Biol. Med., Vol. 42, No. 1, pp. 23–26.

Gomori, G., 1941. The distribution of phosphatase in normal organs and tissues. J. Cell. Comp. Physiol., Vol. 17, No. 1, pp. 71–83.

Gordon, O.L., Beul, E.A., Kortchemykina, K.M., Markova, G.F., and Tarnopolskaya, P.D., 1959. The processes of compensation during deletion of various functions of the gastrointestinal tract (data on total resection of stomach and general resection of the small intestine). XIII Scientific Session of the Institute of Nutrition, Academy of Medical Sciences of the USSR (abstracts of reports). Moscow, p. 86.

Gottlieb, L.S., Robertson, R., and Zamcheck, N., 1962. An electron microscope study of the brush border of the jejunal columnar and crypt epithelium. Am. J. Pathol., Vol. 41, No. 6, pp. 747–751.

Granger, B., and Baker, R.F., 1949. Electron microscope investigation of the striated border of intestinal epithelium. Anat. Record, Vol. 103, No. 3, p. 459.

Granger, B., and Baker, R.F., 1950. Electron microscope investigation of the striated border of intestinal epithelium. Anat. Record, Vol. 107, No. 4, pp. 423–436.

Green, D.E., 1964. On biological membranes. In: Molecular Biology. Problems and Perspectives. In honor of the seventieth birthday of Academician V.A. Engelhardt. Nauka, Moscow, pp. 260-272.

Gryboski, J.D., Thayer, W.R., Gabrielson, I.W., and Spiro, H.M., 1963. Disacchariduria in gastrointestinal disease. Gastroenterology, Vol. 45, No. 5, pp. 633–637.

Gupta, J.D., Dakroury, A.M., and Harper, A.E., 1958. Observations on protein digestion in vivo. I. Rate of disappearance of ingested protein from the gastrointestinal tract. J. Nutr., Vol. 64, No. 3, pp. 447–456.

Halliday, R., 1958. The increase in alkaline phosphatase activity of the duodenum and decrease in absorption of antibodies by the gut induced in young rats by deoxycorticosterone acetate. J. Physiol., Vol. 140, No. 2, pp. 44–45P.

Harrer, C.J., and King, C.G., 1941. Ascorbic acid deficiency and enzyme activity in guinea pig tissues. J. Biol. Chem., Vol. 138, No. 1, pp. 111–121.

Harris, R.S., Chamberlain, J.W., and Benedict, J.H., 1955. Digestion of neutral fats by human subjects. J. Clin. Invest., Vol. 34, No. 5, pp. 685–689.

Hart, S.L., and Nissim, J.A., 1964. First receptor in intestinal sugar transport. The effect of phloridzin and of cetrimide on the intestinal absorption of α-d-glucosamine, *l*-glucose and d-xylose. Nature, Vol. 204, No. 4953, pp. 51–53.

Hartman, R.E., Smith, R.B.W., Hartman, R.S., Butterworth, C.E., and Molesworth, J.M., 1959. The electron microscopy of human intestinal epithelium obtained with the Crosby intestinal biopsy capsule. J. Biophys. Biochem. Cytol., Vol. 5, No. 1, pp. 171–172.

Hartman, R.S., Butterworth, C.E., Hartman, R.E., Crosby, W.H., and Shirai, A., 1960. An electron microscopic investigation of the jejunal epithelium in sprue. Gastroenterology, Vol. 38, No. 4, pp. 506–516.

Hartmann, F., Lenz, H., Lopez-Calleja, C., and Münzenberg, H., 1956. Die Resorptionsweise von Aminosauren aus dem Dünndarm. Arch. Exptl. Pathol. Pharmakol., Vol. 228, No. 4, pp. 403–413.

Hennig, N., Zeitler, G., and Neugebauer, I., 1959. An apparatus for mucous membrane biopsy in a deep section of the intestine. Dtsch. Med. Wschr., Jhrg. 84, pp. 1961–1963.

Herbert, E., 1952. A Study of the Liberation of Orthophosphate from Adenosine Triphosphate by the Stromata of Human Erythrocytes. Philadelphia.

Hewing, E., 1956. Chemical reactions and electric potential in monomolecular layers. In: Monomolecular Layers. IIL, Moscow, pp. 222-235.

Hill, K.J., 1956. Gastric development and antibody transference in the lamb, with some observations on the rat and guinea pig. Quart. J. Exptl. Physiol., Vol. 41, No. 4, pp. 421–432.

Hoffman-Ostenhoff, O., 1959. The origin of enzymes. In: The Origin of Life on Earth (III International Symposium, 19–24 August, 1957). Izd. AN SSSR, Moscow, pp. 205–213.

Hofmann, A.F., 1966. Gastroenterology, Vol. 50, pp. 56–64.
Hofmann, A.F., and Borgström, B., 1963. Hydrolysis of long-chain monoglycerides in micellar solution by pancreatic lipase. Biochim. Biophys. Acta, Vol. 70, No. 3, pp. 317–331.
Holden, G.B.S., 1935. Factors of Evolution. Biomedgiz, Moscow–Leningrad.
Holt, J.H., and Miller, D., 1961. The intestinal brush border as a digestive surface. Localization of aminopeptidase and phosphatase in the mucosal epithelial cell. J. Lab. Clin. Med., Vol. 58, No. 5, p. 827.
Holt, J.H., and Miller, D., 1962. The localization of phosphomonoesterase and aminopeptidase in brush border isolated from intestinal epithelial cells. Biochim. Biophys. Acta, Vol. 58, No. 2, pp. 239–243.
Holzel, A., 1965. Development of intestinal enzyme systems and its relation to diarrhea. Pediat. Clin. N. Am., Vol. 12, No. 3, pp. 635–654.
Honeyman, G. (ed.), 1962. Recent Advances in Chemistry of Cellulose and Starch (a compilation). IIL, Moscow. [English edition: 1960, Heywood and Co.]
Hooft, C., Van Hauwaert, J., De Laey, P., Adriaenssens, K., 1963a. Intestinal lactase deficiency. Lancet, Vol. 2, No. 7311, pp. 791–792.
Hooft, C., Van Hauwaert, J., De Laey, P., Adriaenssens, K., 1963b. Malabsorption after total gastroectomy in childhood. Helv. Pediatr. Acta, Vol. 18, No. 6, p. 502.
Hooper, C.E.S., 1956. Cell turnover in epithelial populations. J. Histochem. Cytochem., Vol. 4, No. 6, pp. 531–540.
Hooper, C.E.S., 1962. Methods for the investigation of cellular renewal in the intestine epithelium. Methods Med. Res., Vol. 9, pp. 326–342.
Iezuitova, N.N., 1963. On hydrolysis of certain poly- and oligosaccharides in the mammalian small intestine. First All-Union Biochemical Conference. Abstracts, Vol. 2, Supplementary (sectional) Conference 1-9. Izd. AN SSSR, Moscow–Leningrad, p. 43.
Iezuitova, N.N., 1965. On the role of membrane digestion in the hydrolysis of poly- and oligosaccharides. Data of scientific conference on the problems "Physiology and pathology of cortico-visceral interrelationships and functional systems of the organism." Ivanovo, pp. 414–417.
Iezuitova, N.N., Ugolev, A.M., and Fedushina, I.N., 1963a. The effect of the rate of perfusion on cavital and membranal hydrolysis of starch and sucrose. Dokl. Akad. Nauk SSSR, Vol. 149, No. 3, pp. 746–749.
Iezuitova, N.N., De Laey, P., Marauska, M., Meboldt, V.R., Timofeeva, N.M., Ugolev, A.M., Fedushina, I.N., and Khalimov, K.I., 1963b. On the significance of membrane digestion in the work of the gastrointestinal tract. In: Physiology and Pathology of Digestive System (reports). Data of the scientific conference in honor of I.P. Rasenkov's seventy-fifth birthday. Moscow, pp. 59–61.
Iezuitova, N.N., De Laey, P., and Ugolev, A.M., 1964a. Digestion of starch in vivo and in vitro in a rat intestine. Biochim. Biophys. Acta, Vol. 86, No. 2, pp. 205–210.
Iezuitova, N.N., Marauska, M.K., and Timofeeva, N.M., 1964b. The effect of the functional state on the course of membrane digestion. First Conference of the All-Union I.P. Pavlov Physiological Society, Vol. 2, Abstracts of Scientific Sessions, No. 1. Moscow–Leningrad, p. 338.
Iezuitova, N.N., Timofeeva, N.M., Koldovskii, O.K., Nurks, Ya. Ya., and Ugolev, A.M., 1964c. Postnatal development of enzymatic activity of the surface of the rat small intestine (invertase, peptidase, lipase). Dokl. Akad. Nauk SSSR, Vol. 154, No. 4, pp. 990–993.
Iezuitova, N.N., Koldovskii, O.K., Nurks, Ya. Ya., Salenietse, I.K., Stepanova, A.T., Timofeeva, N.M., and Ugolev, A.M., 1964d. Concerning the role of cavital and membrane digestion in early postnatal ontogenesis. X Conference of All-Union I.P. Pavlov Physiological Society, Vol. 2, Abstracts of Scientific Sessions, No. 1. Moscow–Leningrad, p. 338.
Iezuitova, N.N., Nadirova, T.Ya., Toropova, N.V., and Ugolev, A.M., 1965a. The characterization of the entry of hexoses into the blood after the administration of poly-, oligo-, and monosaccharides into the gastrointestinal tract. In: Physiology and Pathology of Digestion. Brief communications of scientific conference, 24-28 September, 1965. Lvov, pp. 102–105.

Iezuitova, N.N., Timofeeva, N.M., Koldovskii, O.K., Nurks, Ya. Ya., and Ugolev, A.M., 1965b. On the role of cavital and membrane digestion in early postnatal ontogenesis. In: Realistic Problems of Obstetry and Pediatrics. Scientific Works of Kirghiz Scientific Institute for Care of Motherhood and Childhood, No. 2. Frunze, pp. 114–117.

Iezuitova, N.N., Timofeeva, N.M., Zabelinskii, E.K., and Ugolev, A.M., 1967. Dokl. Akad. Nauk SSSR, Biol. Sci. Sect., Vol. 173, No. 2, pp. 475–478.

Isselbacher, K.J., 1961. Fat absorption and the esterification of fatty acids by intestinal mucosa. Gastroenterology, Vol. 40, No. 2, Part 1, pp. 259–260.

Isselbacher, K.J., 1965. Metabolism and transport of lipid by intestinal mucosa. Feder. Proc., Vol. 24, No. 1, Part I, pp. 16–22.

Isselbacher, K.J., 1966. Gastroenterology, Vol. 50, pp. 78–82.

Isselbacher, K.J., and Senior, J.R., 1964. The intestinal absorption of carbohydrate and fat. Gastroenterology, Vol. 46, No. 3, pp. 287–298.

Ito, S., 1965. J. Cell. Biol., Vol. 27, pp. 475–491.

Izmailov, T.U., 1965. On membranal digestion in forestomachs of cattle. Izv. Akad. Nauk Kaz. SSR, Ser. Biol. Nauk, No. 5, pp. 84–87.

Jackson, S.H., 1952. The effect of food ingestion on intestinal and serum alkaline phosphatase in rats. J. Biol. Chem., Vol. 198, No. 2, pp. 553–559.

Jackson, I.M., 1964. Extrasecretory function of the pancreas and its involvement in the regulation of certain aspects of metabolism. Author's dissertation, Leningrad.

Jacob, F., and Monod, J., 1964. Biochemical and genetic mechanisms of regulation in bacterial cell. In: Molecular Biology. Problems and Perspectives. In honor of the seventieth birthday of Academician V.A. Engelhardt. Nauka, Moscow, pp. 14–39.

Jeffries, G.H., Weser, E., and Sleisenger, M.H., 1964. Malabsorption. Gastroenterology, Vol. 46, No. 4, pp. 434–466.

Johnson, R.B., 1953. A theoretical discussion of ruminant ketosis. Am. J. Vet. Res., Vol. 14, p. 366.

Johnston, C.F., 1967. Science, Vol. 155, pp. 1670–1672.

Johnston, J.M., and Adams, J.F., 1964. Jejunal mucosal appearances after total gastrectomy. Gut, Vol. 5, No. 1, p. 60.

Johnston, J.M., and Wiggans, D.S., 1958. The absorption in vitro of alanyl-phenylalanine. Biochim. Biophys. Acta, Vol. 27, No. 1, p. 224.

Jordan, G., 1934. Manual for Comparative Physiology. Biomedgiz, Moscow–Leningrad.

Josefsson, L., and Sjöstrom, H., 1966. Acta Physiol. Scand., Vol. 67, pp. 27–33.

Ju, J.S., and Nasset, E.S., 1959. Amylase in pancreas, intestine, liver, and serum during fasting, nonprotein diet and realimentation. Proc. Soc. Exptl. Biol. Med., Vol. 100, No. 4, pp. 834–837.

Kadyrov, U.Z., 1961a. Effect of diet poor in lipotropic substances on the production of phospholipids and enzymes by the glands of dog intestine. Vopr. Pitaniya, Vol. 20, No. 2, pp. 16–22.

Kadyrov, U.Z., 1961b. Effect of methionine and choline on the content of phospholipids in the intestinal juice of dogs maintained on a diet poor in lipotropic substances. Vopr. Pitaniya, Vol. 20, No. 4, pp. 15–18.

Kaliteevskaya, T.N., 1957. X-ray observations on the alterations in the gastrointestinal tract during lead poisoning. Sov. Med., No. 4, pp. 96–98.

Kalser, M.H., Roth, J.L.A., Tumen, H., and Johnson, T.A., 1960. Relation of small bowel resection to nutrition in man. Gastroenterology, Vol. 38, No. 4, pp. 605–615.

Kaplan, J.G., and Tacreiter, W., 1966. J. Gen. Physiol., Vol. 50, pp. 9–24.

Kato, Y., and Moog, F., 1958. Difference in response of phosphatases in chick embryo to injection of substrate. Science, Vol. 127, No. 3302, pp. 812–813.

Kauder, E., and Bayless, T.M., 1964. Peroral intestinal biopsy in children: a technique. Am. J. Diseases Children, Vol. 107, No. 6, pp. 582–585.

Kavanau, J.L., 1965. Structure and Function in Biological Membranes, Vol. 1. San Francisco–London–Amsterdam.

Kay, R.E., and Entenman, C., 1959. Effect of whole-body x-irradiation on glucose oxidation by rat small intestine mucosa in vitro. J. Biol. Chem., Vol. 234, No. 7, pp. 1634–1640.

Kazaryan, K.G., and Kulminskaya, A.S., 1966. Dokl. Akad. Nauk SSSR, Vol. 166, pp. 462–465.

Kern, F., Struthers, J.E., and Attwood, W.L., 1963. Lactose intolerance as a cause of steatorrhea in an adult. Gastroenterology, Vol. 45, No. 1, pp. 477–487.

Kerr, R.V. (ed.), 1956. Chemistry and Technology of Starch. Pishchepromizdat, Moscow.

Kessel, D., and Lubin, M., 1963. On the distinction between peptidase activity and peptide transport. Biochim. Biophys. Acta, Vol. 71, No. 3, pp. 656–663.

Kessel, R.G., Prestage, J.J., Sekhon, S.S., Smalley, R.L., and Beams, H.W., 1961. Cytological studies on the intestinal epithelium of *Ascaris lumbricoides* suum. Trans. Am. Microscop. Soc., Vol. 80, No. 1, pp. 103–118.

Khalimov, K.I., 1963. Histochemical study of intestinal content and intestinal epithelium. In: Physiology and Pathology of Digestion and Problems of Resortology and Physiotherapy. Data and abstracts of reports of the conference, 7–13 October, 1963. Gagra, Tbilisi, pp. 176–177.

Kheisin, E.M., 1965. Pynocytosis. In: Manual in Cytology, Vol. 1. Nauka, Moscow–Leningrad, pp. 166–172.

Klotz, A.P., 1964. Intestinal lactase deficiency and diarrhea in adults. Am. J. Digest. Diseases, Vol. 9, No. 5, pp. 345–354.

Klumpp, T.G., and Neale, A.V., 1930. Gastric and duodenal contents of normal infants and children; duodenal enzyme activity and gastric and duodenal reactions (H ions). Am. J. Diseases Children, Vol. 40, pp. 1215–1229.

Knoebel, L.K., and Nasset, E.S., 1957. The digestion and absorption of fat in dog and man. J. Nutr., Vol. 61, No. 3, pp. 405–419.

Knoebel, L.K., and Ryan, J.M., 1965. Effect of exclusion of pancreatic juice on digestion and mucosal absorption of fat in dogs. Proc. Soc. Exptl. Biol. Med., Vol. 118, No. 1, pp. 161–167.

Knowlton, N.P., and Widner, W.P., 1950. The use of x-rays to determine the mitotic and intermitotic time of various mouse tissues. Cancer Res., Vol. 10, No. 1, pp. 59–63.

Kodousek, R., Kojecky, Z., Beatny, J., and Malinsky, J., 1963. Prispevek histochemie a electronove microscopie k problematice whippleovy choroby. Cesk. Gastroenterol. Vyziva, Vol. 17, No. 5, pp. 290–294.

Koldovskii, O., and Chytil, F., 1965. Postnatal development of β-galactosidase activity in the small intestine of the rat. Effect of adrenalectomy and diet. Biochem. J., Vol. 94, No. 1, pp. 266–270.

Koldovskii, O., Danysz, J., Faltová, E., and Hahn, P., 1963. The postnatal proximodistal development of glucose absorption, intestinal alkaline phosphatase activity, and propulsive motility of the intestine in rats. Physiol. Bohemoslov., Vol. 12, No. 3, pp. 208–212.

Koldovskii, O., Heringová, A., Jirsova, A., Chytil, F., and Hosková, J., 1966. Postnatal development of the β-galactosidase activity in the jejunum and ileum of mice, rabbits, and guinea pigs. Cited in: Koldovskii, O., Sunshine, P., and Kretchmer, N., 1966. Gastroenterology, Vol. 50, No. 4, pp. 596–599.

Komissarchik, Ya. Yu., 1965. Cell membrane (plasma membrane). In: Manual of Cytology, Vol. 1. Nauka, Moscow–Leningrad, pp. 153–160.

Koshtoyants, Kh. S., 1950. Fundamentals of Comparative Physiology, Vol. I, 2nd ed. Izd. AN SSSR, Moscow–Leningrad.

Koskowski, W., 1926. The influence of histamine in the intestinal secretion of the dog. J. Pharmacol., Vol. 26, No. 6, pp. 413–419.

Krainick, H.G., and Mohn, G., 1959. Weitere Untersuchungen über den schädlichen Weizenmehleffekt bei der Cöliakie. 2. Die Wirkung der enzymatischen Abbauprodukte des Gliadin. Helv. Pediatr. Acta, Vol. 14, No. 2, pp. 124–140.

Kratzer, F.H., 1944. Amino acid absorption and utilization in the chick. J. Biol. Chem., Vol. 153, No. 1, pp. 237–247.

Kravitskaya, P.S., 1959. Mechanisms of regulation of secretory and motor functions of the stomach at various ages. Dissertation. Alma-Ata.

Kruchekova, A.P., 1939. Amylase and lipase as indicators of changes in the functional state of pancreas in ontogenesis. Phisiologich. J. USSR, Vol. 27, No. 4, pp. 437–444.

Kretchmer, N., 1964. Whither birth defects? Perspectives. Biol. Med., Vol. 8, No. 10, pp. 15–29.

Kuhrt, N.H., Welch, E.A., and Nasset, E.S., 1952. Isolation and identification of monoglycerides in the intestinal contents of humans. J. Am. Oil Chem. Soc., Vol. 29, pp. 271–278.

Kulyukin, S.S., 1964. On the motor-evacuatory function of intestine after total resection of the stomach. Vopr. Onkol., Vol. 10, No. 10, pp. 12–16.

Kursanov, A.L., 1940. Reversible Activity of Enzymes in Living Plant Cells. Izd. AN SSSR, Moscow–Leningrad.

Kurtsin, I.T., 1961. Ionizing Radiation and Digestion. Medgiz, Leningrad.

Kushak, R.I., 1965. Analysis of localization of digestion of oligopeptides in the small intestine by means of thin layer chromatography. Data of II Biochemical Conference of Baltic Republics and Belorussian SSR. Riga, pp. 253–255.

Kushak, R.I., and Ugolev, A.M., 1966. On localization of peptidase activity in the cells of the small intestine of white rats. Dokl. Akad. Nauk SSSR, Vol. 168, No. 2, pp. 477–479.

Kuvaeyeva, I.B., 1962. Effect of diet with low content of indispensable fatty acids and vitamin B_6 on the secretion of phospholipids and enzymes in the intestine. Vopr. Pitaniya, Vol. 21, No. 4, pp. 30–36.

Kuznetsov, N. Ya., 1948. Fundamentals of the Physiology of Insects. Vol. 1. Izd. AN SSSR, Moscow–Leningrad.

Kuznetsova, Z.V., 1963. On the peculiarities of the external secretion of pancreas in pups of various ages. Phisiologich. J. USSR, Vol. 49, No. 2, pp. 242–248.

Larner, J., and Gillespie, R.E., 1956. Gastrointestinal digestion of starch. II. Properties of the intestinal carbohydrases. J. Biol. Chem., Vol. 223, No. 2, pp. 709–726.

Laster, L., and Ingelfinger, F.J., 1961. Intestinal absorption aspects of structure, function, and disease of the small intestine mucosa (concluded). New Engl. J. Med., Vol. 264, No. 2, pp. 1246–1253.

Laws, B.M., and Moore, J.H., 1963. Some observations on the pancreatic amylase and intestinal maltase of the chick. Can. J. Biochem. Physiol., Vol. 41, No. 10, pp. 2107–2121.

Laws, J.W., Shawdon, H., Booth, C.C., and Stewart, J.S., 1963. Correlation of radiological and histological findings in the idiopathic steatorrhea. Brit. J. Med., Vol. 1, p. 1311.

Leblond, C.P., and Messier, B., 1958. Renewal of chief cells and goblet cells in the small intestine as shown by radioautography after injection of thymidine-H^3 into mice. Anat. Record, Vol. 132, No. 3, pp. 247–260.

Leblond, C. P., and Stevens, C.E., 1948. The constant renewal of the intestinal epithelium in the albino rat. Anat. Record, Vol. 100, No. 3, pp. 357–371.

Leblond, C.P., and Walker, B.E., 1956. Renewal of cell populations. Physiol. Reviews, Vol. 36, No. 2, pp. 255–276.

Lee, D.L., 1962. The histochemical localization of leucine aminopeptidase in *Ascaris lumbricoides*. Parasitology, Vol. 52, No. 3–4, pp. 533–538.

Lee, V.V., 1962. On membrane digestion in the chicken intestine. Phisiologich. J. USSR, Vol. 48, No. 12, pp. 1484–1487.

Lehmann, K.E., 1961. An instrument for multiple transoral biopsies of the gastrointestinal tract. Acta Med. Scand., Vol. 161, pp. 205–211.

Leibson, L.G., 1962. Blood Sugar. Regulation of Blood Sugar Content in Animals and Man. Izd. AN SSSR, Moscow–Leningrad.

Lesher, S., Fry, R.J.M., and Kohn, H.J., 1961. Age and the generation time of the mouse duodenal epithelial cell. Exptl. Cell. Res., Vol. 24, No. 2, pp. 334–343.

Leube, W., 1868. Über Verdauungsproduckte des Dünndarmesaffes. Centralbl. Med. Wiss., No. 19, pp. 289–292.

Levenson, S.M., Rosen, H., and Upjohn, H.L., 1959. Nature and appearance of protein digestion products in upper mesenteric blood. Proc. Soc. Exptl. Biol. Med., Vol. 101, No. 1, pp. 178–180.

Levin, R.J., Newey, H., and Smyth, D.H., 1965. The effects of adrenalectomy and fasting on intestinal function in the rat. J. Physiol., Vol. 177, No. 1, pp. 58–73.

Levrat, M., Roche, L., Bret, P., and Anjou, A., 1954. La radiologie intestinale de la colique de plomb. La méga-colon saturnin. Arch. Maladies App. Digest., Vol. 43, No. 2, pp. 206–209.

Lifshitz, F., Klotz, A.P., and Holman, G.H., 1965. Intestinal disaccharidase deficiencies in gluten-sensitive enteropathy. Am. J. Dig. Diseases, Vol. 10, No. 1, pp. 47–57.

Lindemann, B., and Solomon, A.K., 1962. Permeability of luminal surface of intestinal mucosal cells. J. Gen. Physiol., Vol. 45, No. 4, Part I, pp. 801–810.

Lindenbaum, J., 1965. Malabsorption during and after recovery from acute intestinal infection. Brit. Med. J., No. 5457, pp. 326–329.

Lipkin, M., 1965. Cell proliferation in the gastrointestinal tract of man. Feder. Proc., Vol. 24, No. 1, Part I, pp. 10–15.

Littman, A., and Hammond, J., 1965. Diarrhea in adults caused by deficiency in intestinal disaccharidases. Gastroenterology, Vol. 48, No. 3, pp. 237–249.

Livanov, N.A., 1955. Paths of Evolution of Animal World. Analysis of the Organization of the Principal Types of Multicellular Animals. Izd. Sov. Nauka, Moscow.

London, E.S., 1916. Physiology and Pathology of Digestion. Petrograd.

London, E.S., and Lovitskii, Ya. A., 1938. Metabolism in the Organism of Animals and Man. Biomedgiz, Moscow–Leningrad.

Long, C., 1953. Studies involving enzymic phosphorylation. 2. Changes in the hexokinase activity of the small intestine of rats caused by feeding different diets. Biochem. J., Vol. 53, No. 1, pp. 7–12.

Luckey, T.D., 1964. Intestinal flora. In: Conference on Nutrition on Space and Waste Problems. Florida, pp. 227–243.

Lundh, G., 1958. Intestinal digestion and absorption after gastrectomy. Acta Chir. Scand., Suppl. 231.

Lundh, G., 1962. The mechanism of postgastrectomy malabsorption. Gastroenterology, Vol. 42, No. 5, pp. 637–640.

Lüthy, E., and Verzár, F., 1954. Adenosinetriphosphatase and hexokinase in the epithelium of the small intestine in normal and adrenalectomized rats. Biochem. J., Vol. 57, No. 2, pp. 316–320.

MacDonald, W.C., Branborg, L.L., Flick, A.L., Trier, J.S., and Rubin, C.E., 1964a. Studies of celiac sprue. IV. The response of the whole length of the small bowel to a gluten-free diet. Gastroenterology, Vol. 47, No. 6, pp. 573–589.

MacDonald, W.C., Trier, J., and Everett, N.B., 1964b. Cell proliferation and migration in the stomach, duodenum, and rectum of man: radioautographic studies. Gastroenterology, Vol. 46, No. 4, pp. 405–417.

Madanagopalan, N., Shiner, M., and Rowe, B., 1965. Measurements of small intestinal mucosa obtained by peroral biopsy. Am. J. Med., Vol. 38, No. 1, pp. 42–53.

Mali, R., 1886. Chemistry of digestive fluids and digestion. In: L. Herman. Physiology Manual, Vol. 5. St. Petersburg.

Mambeeva, A.A., 1960. Changes in the intramural nerve apparatus of the stomach and intestine in experimental lead poisoning. Izv. Akad. Nauk KazSSR, Ser. Med. Fiziol., No. 2 (14), pp. 66–74.

Mambeeva, A.A., 1963. Motor-evacuatory function of the digestive tract in experimental lead poisoning. Bull. Eksperim. Biol. i Med., Vol. 55, No. 4, pp. 41–44.

Mandels, G.R., 1953. Localization of carbohydrases at the surface of fungus spores by acid treatment. Exptl. Cell Research, Vol. 5, No. 1, pp. 48–55.

Marauska, M.K., Salenietse, I., and Ugolev, A.M., 1963. On the significance of membrane digestion in early postnatal period of ontogenesis of mammals. Data of VI Scientific Conference on Age Morphology, Physiology, and Biochemistry. APN, RSFSR, Moscow, p. 402.

Martin, B.F., and Jacoby, F., 1949. Diffusion phenomenon complicating the histochemical reaction for alkaline phosphatase. J. Anat., Vol. 83, Part 4, pp. 351–363.

Masarei, J., and Simmonds, W.J., 1966. Gut, Vol. 7, pp. 114–118.

Masevich, Ts. G., 1965. The significance of in vivo morphological investigation of the mucosa of duodenum and the small intestine. Data of XIV All-Union Conference of Therapeutists in the Field of "Chronic Diseases of the Intestine (Noninfectious)." Moscow, pp. 39–40.

Masevich, Ts.G., Zabelinskii, E.K., and Ugolev, A.M., 1967. Therapeutic Archive, No. 8, p. 3.

Mashanskii, V.F., 1965. Mitochondria. In: Manual in Cytology. Nauka, Moscow–Leningrad, pp. 200–218.

Mattson, F.H., Benedict, J.H., Martin, J.B., and Beck, L.W., 1952. Intermediates formed during the digestion of triglycerides. J. Nutr., Vol. 48, pp. 335–344.

McGeachin, R.L., and Ford, N.K., 1959. Distribution of amylase in the gastrointestinal tract of the rat. Am. J. Physiol., Vol. 196, No. 5, pp. 972–974.

McLaren, A.D., and Babcock, K.L., 1959. Some characteristics of enzyme reactions at surfaces. In: Subcellular Particles. New York, pp. 23–37.

McLaren, A.D., Mitchell, P.M., and Passow, H. (eds.), 1963. The Cell Interface Reactions. New York.

McMinn, R.M.H., 1954. The rate of renewal of intestinal epithelium in the cat. J. Anat., Vol. 88, Part 4, pp. 527–532.

Menguy, R., 1964. Motor function of the alimentary tract. Ann. Rev. Physiol., Vol. 26, pp. 227–248.

Mikhlin, S. Ya., 1952. A study of the phosphoesterase and phosphoamidase activities of the intestinal juice. Vopr. Pitaniya, Vol. 11, No. 2, pp. 11–12.

Mikhlin, S. Ya., 1955. The content of phosphatases in various sections of digestive tract. Vopr. Med. Khim., Vol. 1, No. 3, p. 210.

Mikhlin, S. Ya., 1962. Effect of experimental folic acid deficiency induced by aminopterin on the secretion of enzymes of the principal digestive glands. Arkhiv Patol., Vol. 24, No. 9, pp. 24–32.

Mikhlin, S. Ya., Nesterin, M.F., and Botchkov, N.P., 1962. Effect of β-acetylpyridine on the enzyme-secreting processes and morphological picture of dog intestine. Patol., Phisiol., i Eksperim. Terapiya, Vol. 6, No. 3, pp. 67–68.

Miller, D., and Crane, R.K., 1961a. The digestive function of the epithelium of the small intestine. I. An intracellular locus of disaccharide and sugar phosphate ester hydrolysis. Biochim. Biophys. Acta, Vol. 52, No. 2, pp. 281–293.

Miller, D., and Crane, R.K., 1961b. The digestive function of the epithelium of the small intestine. II. Localization of disaccharide hydrolysis in the isolated brush border portion of intestinal epithelial cells. Biochim. Biophys. Acta, Vol. 52, No. 2, pp. 293–298.

Miller, D., and Crane, R.K., 1963. The digestion of carbohydrates in the small intestine. Am. J. Clin. Nutr., Vol. 12, No. 3, pp. 220–227.

Millington, P.F., and Finean, J.B., 1962. Electron microscope studies of the structure of the microvilli on principal epithelial cells of rat jejunum after treatment in hypo- and hypertonic saline. J. Cell. Biol., Vol. 14, No. 1, pp. 125–139.

Miner, R.W. (ed.), 1954. Properties of surfaces. Ann. New York Acad. Sci., Vol. 58, Part 6, pp. 721–970.

Mitz, M.A., and Summaria, L.J., 1961. Synthesis of biologically active cellulose derivatives of enzymes. Nature, Vol. 189, No. 4764, pp. 576–577.

Mohiuddin, A., 1962. Cytoplasmic vesicles in absorptive cells lining the gut. Acta Anat., Vol. 50, Part 1, pp. 103–111.

Mohiuddin, A., 1964. The jejunal villi of the adult white rat during fat absorption. Acta Anat., Vol. 59, No. 3, pp. 201–211.

Moog, F., 1965. Biochemistry of Animal Development, Vol. 1. Academic Press, London, pp. 307-365.

Moog, F., and Ford, E., 1957. Influence of exogenous ACTH of body weight, adrenal growth duodenal phosphatase, and liver glycogen in the chick embryo. Anat. Record, Vol. 128, No. 3, p. 592.

Morrison, W.L., and Neurath, H., 1953. Proteolytic enzymes of the formed elements of human blood. I. Erythrocytes. J. Biol. Chem., Vol. 200, No. 1, pp. 39–51.

Mosinger, B., Placer, Z., and Koldovskii, O., 1959. Passage of insulin through the wall of the gastrointestinal tract of the infant rat. Nature, Vol. 184, No. 4694, pp. 1245–1246.

Moss, W.T., 1957. The effect of irradiating the exteriorized small bowel on sugar absorption. Am. J. Roentgenol., Vol. 78, No. 5, pp. 851–854.

Müller, M., Röhlich, P., Toth, J., and Törö, I., 1963. Fine structure and enzymic activity of protozoan food vacuoles. Ciba Found. Symp. Lysosomes, pp. 201–216.

Myrbäck, K., and Ortenblad, B., 1936. Trehalose und Hefe, I. Biochem. Z., Vol. 288, No. 5–6, pp. 239–337.

Myrbäck, K., and Ortenblad, B., 1937. Trehalose und Hefe, II. Mitt. Trehalasewirkung von Hefepräparaten. Biochem. Z., Vol. 291, No. 1–3, pp. 61–69.

Myrbäck, K., and Vasseur, E., 1943. Über die Lactosegärung und die Lokalisation der Enzyme in der Hefezelle. Hoppe-Seyler's Z. Physiol. Chemie, Vol. 277, No. 4–6, pp. 171–180.

Nachlas, M.M., Monis, B., Rosenblatt, D., and Seligman, A.M., 1960. Improvement in the histochemical localization of leucine aminopeptidase with a new substrate, L-leucyl-4-methoxy-2-naphthylamide. J. Biophys. Biochem. Cytol., Vol. 7, No. 2, pp. 261–264.

Nadirova, T. Ya., 1965. Pathological and compensatory reactions of membrane digestion of carbohydrates in lead poisoning. In: Physiology and Pathology of Digestion. Brief account in reports of scientific conference, 24-28 September, 1965. Lvov, pp. 192–194.

Nadirova, T. Ya., and Ugolev, A.M., 1967. Bull. Exp. Biol. Med. (USSR) (English Transl.), No. 11.

Nadirova, T.Ya., Ovdeichuk, R.A., and Ugolev, A.M., 1965a. Certain adaptive and compensatory changes in membrane digestion in rats. In: Problems of Biochemical Adaptation. Symposium, Moscow, pp. 48–49.

Nadirova, T. Ya., Timofeeva, V.A., and Ugolev, A.M., 1965b. On the localization of invertase activity in the cells of the small intestine of white rats. Bull. Eksperim. Biol. i Med., Vol. 49, No. 3, pp. 29–33.

Nasonov, D.N., 1959. Local Reaction of Protoplasm and Propagating Stimulation. Izd. AN SSSR, Moscow–Leningrad.

Nasset, E.S., 1962. Amino acids in gut contents during digestion in the dog. J. Nutr., Vol. 76, No. 2, Part I, p. 131.

Nasset, E.S., and Jin Soon Ju, 1961. Admixture of endogenous protein with ingested protein in digestive tracts of dogs and rats. Feder. Proc., Vol. 20, No. 1, Part I, p. 245.

Nasset, E.S., Schwartz, P., and Weiss, H.V., 1955. The digestion of proteins in vivo. J. Nutr., Vol. 56, No. 3, pp. 83–94.

Neifakh, S.A., 1964. Catalytic functions of actomyosin and membrane transport. In: Molecular Biology. Problems and Perspectives. In honor of the seventieth birthday of Academician V.A. Engelhardt. Nauka, Moscow, pp. 273–281.

Neifakh, S.A., and Vasilets, I.M., 1964. Actomyosinlike protein in the outer membrane of liver cells. Vopr. Med. Khim., Vol. 10, No. 3, pp. 326–328.

Nelson, N., 1944. A photometric adaptation of the Somogyi method for the determination of glucose. J. Biol. Chem., Vol. 153, No. 3, pp. 375–380.

Newey, H., and Smyth, D.H., 1957. Intestinal absorption of dipeptides. J. Physiol., Vol. 135, No. 2, pp. 43–44.

Newey, H., and Smyth, D.H., 1959. The intestinal absorption of some dipeptides. J. Physiol., Vol. 145, No. 1, pp. 48–56.

Newey, H., and Smyth, D.H., 1960a. Intracellular hydrolysis of dipeptides during intestinal absorption. J. Physiol., Vol. 152, No. 2, pp. 367–380.

Newey, H., and Smyth, D.H., 1960b. Absorption rats of glycine and glycyl-glycine. J. Physiol., Vol. 152, No. 2, pp. 70–71.

Newey, H., and Smyth, D.H., 1961. Two-stages transfer of glycine by the intestine in vitro. J. Physiol., Vol. 157, No. 1, pp. 15–16.

Newey, H., and Smyth, D.H., 1963. Specificity of carriers in intestinal transfer of glycine. J. Physiol., Vol. 165, No. 2, p. 74.

Newey, H., Parsons, B.J., and Smyth, D.H., 1959. The site of action of phloridzin in inhibiting intestinal absorption of glucose. J. Physiol., Vol. 148, No. 1, pp. 83–92.

Newey, H., Sanford, P.A., and Smyth, D.H., 1962. The functional topography of the intestinal epithelial cell in carbohydrate absorption. J. Physiol., Vol. 161, No. 2, p. 52.

Newey, H., Sanford, P.A., and Smyth, D.H., 1963. Location of function in the intestinal epithelial cell in relation to carbohydrate absorption. J. Physiol., Vol. 168, No. 2, pp. 423–434.

Nissim, J.A., 1964. Mechanism of intestinal absorption. The concept of a spectrum in intracellular plasma. Nature, Vol. 204, No. 4954, pp. 148–151.

Northrop, J.H., Kunitz, R., and Herriot, R., 1950. Crystalline Enzymes. IIL, Moscow. [English edition: 1948, Columbia University Press.]

Novikoff, A.B., 1961. Lysosomes and related particles. In: J. Brachet and A.E. Mirsky. The Cell, Vol. 2. New York–London.

Obel, E., 1959. On the transition from anerobiosis to aerobiosis. In: Origin of Life on Earth (III International Symposium, 19–24 August, 1957). Izd. AN SSSR, Moscow, pp. 583–590.

Oparin, A., 1959. Biochemical processes in simplest structures. In: Origin of Life on Earth (III International Symposium, 19–24 August, 1957). Izd. AN SSSR, Moscow, pp. 433–440.

Oppenheimer, K., and Kuhn, R., 1932. Enzymes. ONTI, Moscow–Leningrad.

Overton, J., and Shoup, J., 1964. Fine structure of cell surface specializations in the maturing duodenal mucosa of the chick. J. Cell. Biol., Vol. 21, No. 1, pp. 75–85.

Padykula, H.A., 1962. Recent functional interpretations of intestinal morphology. Feder. Proc., Vol. 21, No. 6, pp. 873–879.

Padykula, H.A., Strauss, E.W., Ladman, A.J., and Gardner, F.H., 1961. A morphologic and histochemical analysis of the human jejunal epithelium in nontropical sprue. Gastroenterology, Vol. 40, No. 6, pp. 735–765.

Palay, S.L., and Karlin, L.J., 1959a. An electron microscopic study of the intestinal villus. I. The fasting animal. J. Biophys. Biochem. Cytol., Vol. 5, No. 3, pp. 363–372.

Palay, S.L., and Karlin, L.J., 1959b. An electron microscopic study of the intestinal villus. II. The pathway of fat absorption. J. Biophys. Biochem. Cytol., Vol. 5, No. 3, pp. 373–384.

Palmer, E.D., 1963. Clinical Gastroenterology. New York–London.

Pander, A., Semenza, G., and Auricchio, S., 1963. Intestinal Absorption und Malabsorption der Disaccharide. Schweiz. Med. Wschr., Vol. 93, p. 1272.

Parshin, A.N., and Rubel, L.N., 1951. The nature of products of protein digestion, which are absorbed from the intestine into the blood. Dokl. Akad. Nauk SSSR, New Series, Vol. 77, No. 2, pp. 313–315.

Parsons, D.S., 1963. Quantitative aspects of pinocytosis in relation to intestinal absorption. Nature, Vol. 199, No. 4899, pp. 1192–1193.

Parsons, D.S., and Prichard, J.S., 1965. Hydrolysis of disaccharides during absorption by perfused small intestine of *amphibia*. Nature, Vol. 208, No. 5015, pp. 1097–1098.

Patt, H.M., and Quastler, H., 1963. Radiation effects on cell renewal and related systems. Physiol. Reviews, Vol. 43, No. 3, pp. 357–396.

Patzelt, V., 1936. Der Darm. In: W. von Möllendorf. Handbuch der mikroskopischen Anatomie des Menschen, Vol. 5. Berlin.

Pavlov, I.P., 1897. Lectures on the work of principal digestive glands. Complete Works, Vol. 2, Book 2. Izd. AN SSSR, Moscow–Leningrad, 1951.

Pchelin, V.A., 1951. Surface Properties of Proteins. (Two-Phase State of Proteins). State scientific-technical publications of light industry, Moscow.

Perepelkin, S.R., 1960. Alterations in Secretion and Excretion of Stomach and Small Intestine in Poisoning by Radioactive Substances. Medgiz, Moscow.

Peternel, W.W., 1965. Lactose tolerance in remission to intestinal activity. Gastroenterology, Vol. 48, No. 3, pp. 299–306.

Pierce, H.B., Nasset, E.S., and Murlin, J.R., 1935. Enzyme production in a transplanted loop of the upper jejunum. J. Biol. Chem., Vol. 108, No. 1, pp. 239–250.

Pilson, M.E.Q., and Kelly, A.L., 1962. Composition of the milk from *Zalophus californianus*, the California sea lion. Science, Vol. 135, No. 3498, pp. 104–105.

Pimparkar, B.D., 1964. In: H.L. Bockus. Gastroenterology, Vol. 2. Philadelphia–London.

Pirie, N.W., 1959. Chemical variety and the problem of origin of life. In: Origin of Life on Earth (III International Symposium, 19–24 August, 1957). Izd. AN SSSR, Moscow, pp. 79–87.

Platt, B.S., 1961. Digestion in infancy. Feder. Proc., Vol. 20, No. 1, Part III, Suppl. No. 7, pp. 188–195.

Plosscowe, R.P., Berg, G.G., and Segal, H.L., 1963. Enzyme histochemical studies of human gastric and jejunal biopsy specimens in normal and disease states. Am. J. Digest. Diseases, Vol. 8, No. 4, pp. 311–318.

Plotkin, G.R., and Isselbacher, K.J., 1964. The enzymatic demonstration of disaccharidase deficiency in the intestinal mucosa of nontropical sprue and other malabsorption states. Gastroenterology, Vol. 46, No. 6, p. 756.

Poglazov, B.F., 1965. The Structure and Function of Contractile Proteins. Nauka, Moscow.

Ponz, F., 1963/1964. The active transport of sugars through the intestine, and its relations with the sodium ion. Arquiv. Port. Bioquim., Vol. 7, pp. 1–25, 93.

Pope, J.L., McPherson, J.S., and Tudwell, H.S., 1966. J. Biol. Chem., Vol. 245, pp. 2306–2310.

Prader, A., and Auricchio, S., 1965. Defect of intestinal disaccharide absorption. Ann. Rev. Med., Vol. 16, pp. 345–358.

Przheletska, A., Ejsmond, G., Sarzala, E.G., and Taracha, M., 1961. Cytochemical study of the activity of alkaline phosphatase in the mucosa of mouse small intestine. V International Congress of Biochemistry. Abstracts, Vol. 1, Section 1–13. Izd. AN SSSR, Moscow, pp. 399–400.

Quastler, H., 1963. Effects of irradiation on intestinal mucosal cell population. Feder. Proc., Vol. 22, No. 6, Part I, pp. 1330–1333.

Quastler, H., and Hampton, J.C., 1962. Effects of ionizing radiation on the fine structure and function of the intestinal epithelium of the mouse. Rad. Res., Vol. 17, No. 6, pp. 914–931.

Rahn, O., and Leet, M., 1949. Adaptive enzymes induced by insoluble substrates. J. Bacteriol., Vol. 58, No. 5, pp. 714–715.

Rajan, K.T., Rao, P.S.S., Ponnusamy, J., and Baker, S.J., 1961. The oral glucose-tolerance test in tropical malabsorption syndrome. Brit. Med. J., No. 5218, pp. 29–31.

Rakhimov, K.R., 1965. On influence of high temperature and insulation on the level of cavital and membrane digestion in the intestine. In: Physiology and Pathology of Digestion. Abstracts of reports of scientific conference, 24–28 September, 1965. Lvov, pp. 247–248.

Raskin, H.F., Wenger, J., Sklar, M., Pleticka, S., and Yarema, W., 1958. The diagnosis of cancer of the pancreas, biliary tract, and duodenum by combined cytologic and secretory methods. I. Exfoliative cytology and a description of a rapid method of duodenal intubation. Gastroenterology, Vol. 34, No. 6, pp. 996–1008.

Ravin, H.A., and Seligman, A.M., 1953. Determinants of the specificity of action of pancreatic lipase. Arch. Biochem. Biophys., Vol. 42, No. 2, pp. 337–354.

Razenkov, I.P., 1948. The Role of the Gastrointestinal Tract in Intermediary Metabolism. Izd. AN SSSR, Moscow.

Roberson, R.H., and Schaible, P.J., 1958. Zinc requirement of the chick. Science, Vol. 127, No. 3303, pp. 875–876.

Robinson, G.B., and Shaw, B., 1960. The hydrolysis of dipeptides by different regions of rat small intestine. Biochem. J., Vol. 77, No. 2, pp. 351–356.

Roginskii, S.Z., 1948. Adsorption and Catalysis on Heterogeneous Surfaces. Izd. AN SSSR, Moscow–Leningrad.

Röhmann, F., and Nagano, J., 1903. Über die Resorption und die fermentative Spaltung des Disaccharide im Dünndarm des ausgewachsenen Hundes. Pflüg. Arch. Ges. Physiol., Vol. 95, No. 11–12, pp. 533–605.

Rokos, J., Hahn, P., Koldovskii, O., and Procházka, P., 1962a. Postnatalni vyvoj aktivity pankreaticke lipazy a vliv kortizonu na tento vyvoj u krys. Cesk. Fysiol., Vol. 11, No. 3, pp. 210–211.

Rokos, J., Hahn, P., Koldovskii, O., and Procházka, P., 1962b. Postnatalni vyvoj aktivity lipasy steny tenkeho streva. Cesk. Fysiol., Vol. 11, No. 5, p. 472.

Rosenberg, T., and Vilbrandt, V., 1955. The role of enzymatic processes in the permeability of the intestinal membrane. In: Modern Problems of Cytology. IIL, Moscow, pp. 183–212.

Ross, J.R., and Moore, V.A., 1961. Small intestinal biopsy capsule utilizing hydrostatic and suction principles. Gastroenterology, Vol. 40, No. 1, pp. 113–119.

Ross, J.R., and Moore, V.A., 1963. Peroral small-intestine biopsy: technique and application. In: Current Gastroenterology. New York.

Rothstein, A., 1954. The enzymology of the cell surface. Protoplasmatologia. Handbuch der Protoplasmaforsch., Vol. 2. Cytoplasma, E4, pp. 1–86.

Rothstein, A., and Meier, R., 1948. The relationship of the cell surface to metabolism. I. Phosphatases in the cell surface of living yeast cells. J. Cell. Comp. Physiol., Vol. 32, No. 1, pp. 77–95.

Rothstein, A., Meier, R.C., and Scharff, T.G., 1953. Relationship of cell surface to metabolism. IX. Digestion of phosphorylated compounds by enzymes located on surfaces of intestinal cell. Am. J. Physiol., Vol. 173, No. 1, pp. 41–46.

Rowlands, E.N., 1961. Clinical implication of recent advances in the physiology of motility and absorption. Proc. Roy. Soc. Med., Vol. 54, No. 9, pp. 775–777.

Rubin, C., and Dobbins, W., 1965. Peroral biopsy of the small intestine. A review of its diagnostic usefulness. Gastroenterology, Vol. 49, No. 6, pp. 676–697.

Rubin, C.E., Brandborg, L.L., Phelps, P.C., and Taylor, H.C., 1960. Studies of celiac disease. I. The apparent identical and specific nature of the duodenal and proximal jejunal lesion in celiac disease and idiopathic sprue. Gastroenterology, Vol. 38, No. 1, pp. 28–49.

Rubin, C.E., Brandborg, L.L., Flick, A.L., MacDonald, W.C., Parkins, R,A., Parmentier, C.M., Phelps, P.C., Sribhibhadh, S., and Trier, J.S., 1962a. Biopsy studies on the pathogenesis of celiac sprue. In: Ciba Found. Study Group Intestinal Biopsy, No. 14. Boston, p. 67.

Rubin, C.E., Brandborg, L.L., Flick, A.L., Phelps, P.C., Parmentier, C.M., and van Niel, S., 1962b. Studies of celiac sprue. III. The effect of repeated wheat instillation into the proximal ileum of patients on gluten-free diet. Gastroenterology, Vol. 43, No. 6, pp. 621–641.

Rubin, G.E., 1966. Gastroenterology, Vol. 50, pp. 65–77.

Rumyantsev, P.P., 1965. Specialized structures of cell surface. In: Manual in Cytology, Vol. 1. Nauka, Moscow–Leningrad, pp. 160–166.

Ruttloff, H., Friese, R., and Täufel, K., 1965. Zur Spaltung und Resorption von Oligosacchariden im Dünndarm. II. Bildung von Glucose in Gewebschnitten des Rattendarmes aus Maltooligosacchariden. Hoppe Seyler's Z. Physiol. Chemie, Vol. 341, No. 1–3, pp. 134–142.

Ruttloff, H., Noack, R., Friese, R., and Schenk, G., 1964. Zur Lokalisation von Carbohydrasen im Burstensaum der Rattenmucosa. Biochem. Z., Vol. 341, No. 1, pp. 15–22.

Rybak, B.J., 1962. Electron microscopic studies of intestinal lesions. I. Aminopterin-induced lesions in mice. Gastroenterology, Vol. 42, No. 3, pp. 306–318.

Ryss, S.M., 1965. Certain aspects of the problem of nonspecific ulcerated colitis. Data of XIV All-Union Conference of Therapists on the Problem of "Chronic Diseases of the Stomach (Noninfectious)." Moscow, pp. 54–55.

Sakula, J., and Shiner, M., 1957. Celiac disease with atrophy of the small-intestine mucosa. Lancet, Vol. II, No. 7001, pp. 876–877.

Salem, S.N., and Truelove, S.C., 1965. Small-intestinal and gastric abnormalities in ulcerative colitis. Brit. Med. J., No. 5438, pp. 827–831.

Salenietse, I.K., Stepanova, A.T., and Ugolev, A.M., 1963. Certain regularities in the development of digestive functions in ontogenesis. In: Physiology and Pathology of Digestion of Problems of Kurortology and Physiotherapy. Data and abstracts of reports of conference 7–13 October, 1963. Gagra, Tbilisi, pp. 151–152.

Samloff, M., Davis, J., and Schenk, E.A., 1965. A clinical and histochemical study of celiac disease before and during a gluten-free diet. Gastroenterology, Vol. 48, No. 2, pp. 155–172.

Santini, R., Aviles, J., and Sheehy, T.W., 1960. Sucrase activity in the intestinal mucosa of patients with sprue and normal subjects. Am. J. Digest. Diseases, Vol. 5, No. 12, pp. 1059–1062.

Sarda, L., and Desnuelle, P., 1958. Action de la lipase pancréatique sur les esters en émulsion. Biochim. Biophys. Acta, Vol. 30, No. 3, pp. 513–521.

Saunders, S.J., and Isselbacher, K.J., 1966. Intestinal absorption of amino acids. Gastroenterology, Vol. 50, No. 4, pp. 586–595.

Savary, P., and Desnuelle, P., 1956. Sur quelques éléments de spécificité pendant l'hydrolyse enzymatique des triglycérides. Biochim. Biophys. Acta, Vol. 21, No. 2, pp. 349–360.

Schedl, H.P., and Clifton, J.A., 1963. Solute and water absorption by the human small intestine. Nature, Vol. 199, No. 4900, pp. 1264–1267.

Schiebler, Th., 1955. Neuere morphologische Anschauungen über die Darmresorption. Dtsch. med. Wschr., Vol. 80, No. 51, pp. 1885–1887.

Schmidt, W., 1961. Elektronmikroskopische Untersuchung des intrazellularen Stofftransportes in der Dünndarmepithelzelle nach Markierung mit Myofer. Z. Zellforsch., Vol. 54, No. 6, pp. 803–806.

Semenza, G., Tosi, R., Valloton-Delachaux, M.C., and Mülhaupt, E., 1964. Sodium activation of human intestinal sucrose and its possible significance in the enzymic organization of brush border. Biochim. Biophys. Acta, Vol. 89, No. 1, pp. 109–116.

Sheehy, T.W., 1964. The Small Intestine. Its Functions and Diseases. New York.

Sheehy, T.W., and Anderson, P.R., 1965. Disaccharidase activity in normal and diseased small bowel. Lancet, Vol. II, No. 7401, pp. 1–4.

Sheehy, T.W., Cohen, W.H., and Brodsky, J.P., 1963. The intestinal lesion in the initial phase of tropical (military) sprue. Am. J. Digest. Diseases, Vol. 8, No. 10, pp. 826–836.

Sheehy, T.W., Artenstein, M.S., and Green, R.W., 1964. Small intestinal mucosa in certain viral diseases. J. Am. Med. Assoc., Vol. 190, No. 12, pp. 1023–1028.

Sheludko, A., 1960. Textbook of Colloid Chemistry. IIL, Moscow. [English edition: American Elsevier.]
Shestopalova, N.M., 1959. The regeneration of mucosa of the small intestine of mammals and birds after experimental injury. In: Problems of Regeneration and Cell Division (the works of conference on the problems of regeneration and cell multiplication). Moscow, pp. 154–158.
Shestopalova, N.M., Avakyan, A.A., and Reingold, V.N., 1961. Electron microscopic comparative study of the structure of the epithelium of duodenum of mammals and amphibious animals. Tsitologiya, Vol. 3, No. 2, pp. 125–136.
Shestopalova, N.M., Avakyan, A.A., Reingold, V.N., and Tkal, V.V., 1960. Electron microscopic study of plastic processes of intestinal epithelium of certain mammals. Arkhiv. Anat., Gistol. i Embriol., Vol. 38, No. 3, pp. 34–44.
Shiner, M., 1956a. Duodenal biopsy. Lancet, Vol. I, No. 6906, pp. 17-19.
Shiner, M., 1956b. Jejunal-biopsy tube. Lancet, Vol. I, No. 6907, p. 85.
Shiner, M., 1957. Duodenal and jejunal biopsies. I. A discussion of the method, its difficulties, and applications. Gastroenterology, Vol. 33, No. 1, pp. 64–70.
Shiner, M., Lacy, D., and Hudson, R.H., 1962. Electron microscope study of fat absorption in normal subjects and in patients with idiopathic steatorrhea. In: Ciba Found. Study Group on Intestinal Biopsy, No. 14. Boston, p. 24.
Shiskov, V.M., 1966. Veterinariya, Vol. 4, p. 78.
Shishova, O.A., Ogurtsova, L.A., and Kasatochkin, V.I., 1961. The kinetics of absorption of amino acids in the intestine. Phisiologich. J. USSR, Vol. 47, No. 5, pp. 630–637.
Shishova-Kasatochkina, O.A., 1964a. Active transport of amino acids in the intestine. Data of the symposium "Physiology and Pathology of Absorption in the Gastrointestinal Tract." Odessa, pp. 80–102.
Shishova-Kasatochkina, O.A., 1964b. Certain problems of absorption of amino acids in the intestine. Vopr. Pitaniya, Vol. 23, No. 6, pp. 3–17.
Shlygin, G.K., 1952. Secretion of intestinal enzymes. Usp. Sovrem. Biol., Vol. 33, No. 1, pp. 14–32.
Shlygin, G.K., 1958. On the specific enzymatic adaptation of the digestive tract. Vopr. Pitaniya, Vol. 17, No. 2, pp. 3–14.
Shlygin, G.K., 1964. The fundamentals of the enzymatic processes in the intestine. Vestn. Akad. Med. Nauk, No. 5, pp. 21–31.
Shtraub, F.B., 1963. Biokhimiya, Budapest.
Shtraub, F.B., and Sabolchi, G., 1964. On the dynamic aspects of the structure of enzymes. In: Molecular Biology. Problems and Perspectives. In honor of the seventieth birthday of Academician V.A. Engelhardt. Nauka, Moscow, pp. 182–187.
Shvetsova, O.I., 1964. Utilization of nutrients in dogs with excluded pancreatic secretion under qualitatively different nourishment. Vopr. Pitaniya, Vol. 23, No. 3, p. 58.
Sjöstrand, F.S., 1956. The ultrastructure of cells as revealed by the electron microscope. In: International Review of Cytology, Vol. 5. New York, pp. 455–533.
Sjöstrand, F.S., 1963. The ultrastructure of the plasma membrane of columnar epithelium cells of the mouse intestine. J. Ultrastruct. Res., Vol. 8, No. 5–6, pp. 517–541.
Sklyarov, Ya. P., and Yaremko, E.E., 1964. The role of membrane concentration of substances in absorption in the small intestine. Phisiologich. J. USSR, Vol. 50, No. 10, pp. 1289–1295.
Sleisenger, M.H., 1961. Clinical and metabolic studies in nontropical sprue, New Engl. J. Med., Vol. 265, pp. 49–56.
Smith, C.A., 1959. The Physiology of the Newborn Infant. Springfield.
Smyth, D.H., 1961. Intestinal absorption. Proc. Roy. Soc. Med., Vol. 54, No. 9, pp. 769–773.
Spencer, R.P., 1960. The Intestinal Tract. Structure, Function, and Pathology in Terms of the Basic Sciences. Springfield.
Spencer, R.P., 1964a. Spatial distribution of intestinal activities. Yale J. Biol. Med., Vol. 36, No. 4, pp. 279–294.
Spencer, R.P., 1964b. Variation of intestinal activity with age: a review. Yale J. Biol. Med., Vol. 37, No. 2, pp. 105–129.
Spencer, R.P., and Knox, W.E., 1960. Comparative enzyme apparatus of the gut mucosa. Feder. Proc., Vol. 19, No. 4, Part I, pp. 886–897.

Starling, E.H., 1906. Mercers' Company Lectures on Recent Advances in the Physiology of Digestion. London.

Stefenson, M., 1951. Metabolism of Bacteria. IIL, Moscow.

Stepanova, A.T., 1960. Comparative ontogenesis of the enzymatic function of pancreas. VI. Conference of young scientists, Institute of Normal and Pathological Physiology. Academy of Medical Sciences of the USSR, Abstracts of reports. Moscow, pp. 81–83.

Stepanova, A.T., 1966. The development and adaptation of the enzymatic (amylolytic) function of the pancreas in ontogenesis of certain birds and mammals. Dissertation. Leningrad.

Sterk, N.N., and Kretchmer, N., 1964. Studies of small intestine during development. IV Digestion of lactose as related to lactosuria in the rabbit. Pediatrics, Vol. 34, No. 5, pp. 609–614.

Strelin, G.S., and Galkovskaya, K.F., 1964. On the significance of the size of the dose in the reaction changes of the mucosa in mouse intestine during irradiation with Co^{60}. Radiobiologiya, Vol. 4, No. 5, pp. 685–692.

Sunshine, P., and Kretchmer, N., 1964. Intestinal disaccharidases. Absence in two species of sea lions. Science, Vol. 144, No. 3620, pp. 850–851.

Surinov, B.P., 1965. High-activity water-insoluble compounds of trypsin and chymotrypsin with cellulose. Dissertation. Leningrad Chemico-Pharmaceutical Institute, Leningrad.

Sutton, D.D., and Lampen, J.O., 1962. Localization of sucrose and maltose fermenting systems in Saccharomyces cerevisiae. Biochim. Biophys. Acta, Vol. 56, No. 2, pp. 303–312.

Swanson, M.A., 1948. Studies on the structure of polysaccharides. IV. Relation of the iodine color to the structure. J. Biol. Chem., Vol. 172, No. 2, pp. 825–837.

Swanson, V., and Thomassen, R., 1965. Pathology of the jejunum mucosa in tropical sprue. Am. J. Pathol., Vol. 46, No. 4, pp. 511–552.

Tashev, T., and Angelova, Z., 1963. Histochemical studies of the activity of lipase, alkaline, and acid phosphatase in the digestive tract of rats subjected to sublethal doses of x-rays on various dietary regimens. Vopr. Pitaniya, Vol. 22, No. 4, pp. 12–19.

Täufel, K., 1965. Zur Perspektive der Enzymologie bei Nahrung und Ernährung. Die Nahrung, Vol. 9, No. 3, pp. 265–285.

Tidball, C.S., 1964. Magnesium and calcium as regulators of intestinal permeability. Am. J. Physiol., Vol. 206, No. 1, pp. 243–246.

Timofeeva, N.M., 1963. Localization of the hydrolysis of certain dipeptides in the small intestine of white rats. First All-Union Biochemical Conference, Abstracts, Vol. II. Supplementary (sectional) meeting, 1–9. Izd. AN SSSR, Moscow–Leningrad, p. 50.

Timofeeva, N.M., 1965. Concerning the mechanism of hydrolysis of peptides in the small intestine. Data on the scientific conference on the problem "Physiology and Pathology of Cortico–Visceral Interrelationships and Functional Systems of the Organism," Vol. 2. Ivanovo, pp. 343–345.

Timofeeva, N.M., Iezuitova, N.N., Nadirova, T.Ya., and Ugolev, A.M., 1965. Characterization of digestive functions of intestinal epithelium in severe irradiations. In: Physiology and Pathology of Digestion. Brief account in reports of the scientific conference, 24–28 September, 1965. Lvov, pp. 272–274.

Tissières, A., 1948a. L'activité des phosphomonoestérases et des pyrophosphatases dans le rein et l'intestin du rat surrénalectomisé et l'action du desoxycorticosterone. Acta Anat., Vol. 5, No. 3, pp. 224–234.

Tissières, A., 1948b. L'influence de la castration, du testostérone et de l'oestradiol sur les phosphatases du rein chez le rat. Acta Anat., Vol. 5, No. 2, pp. 235–242.

Tokin, I.B., 1959. The ultrastructure of the brush of the intestinal cells of *Parascaris equorum*. Dokl. Akad. Nauk SSSR, Vol. 125, No. 4, pp. 902–904.

Toropova, N.V., 1964. On the cleavage of starch in the small intestine of embryos and newborn rabbits. In: The Realistic Problems of Obstetry and Pediatry. Collective scientific works of Kirghiz Scientific Research Institute for the Care of Mothers and Children, Vol. 2. Frunze, pp. 158–159.

Toropova, N.V., Toktorbayeva, O.T., and Toropov, Yu.M., 1965a. On the Amylolytic Activity in the Embryos and Newborn. Reports of the seventh scientific conference on the problem of age morphology, physiology, and biochemistry. Moscow, pp. 464–465.

Toropova, N.V., Toropov, Yu. M., and Toktorbayeva, O.T., 1965b. On the relationship of the processes of cavital and membrane digestion in animal ontogenesis. In: Metabolism in Animals and Plants. Frunze, pp. 162–178.

Triantaphyllopoulos, E., and Tuba, J., 1959. Changes in intestinal and serum alkaline phosphatase levels during absorption of certain amino acids. Can. J. Biochem. Physiol., Vol. 37, No. 5, pp. 711–719.

Trier, J.S., 1962. Morphologic alterations induced by methotrexate in the mucosa of human proximal intestine. I. Serial observations by light microscopy. Gastroenterology, Vol. 42, No. 3, pp. 295–305.

Trier, J., and Rubin, C., 1964. Electron microscopy of the gut: a word of caution. Gastroenterology, Vol. 47, No. 3, pp. 313–315.

Trier, J., and Rubin C., 1965. Electron microscopy of the small intestine: a review. Gastroenterology, Vol. 49, No. 5, pp. 574–603.

Trier, J.S., Phelps, P.C., and Rubin, C.E., 1963. Electron microscopy of mucosa of small intestine. J. Am. Med. Assoc., Vol. 183, No. 9, pp. 768–774.

Troshin, A.S., 1956. Problem of Cell Permeability. Izd. AN SSSR, Moscow–Leningrad.

Tuba, J., and Dickie, N., 1955. The role of alkaline phosphatase in intestinal absorption. IV. The effect of various proteins on levels of the enzyme in intestinal mucosa. Can. J. Biochem. Physiol., Vol. 33, No. 1, pp. 89–92.

Tuba, J., and Robinson, M.I., 1953. The response of intestinal alkaline phosphatase of fasted rats to forced feeding of fat. J. Biol. Chem., Vol. 203, No. 2, pp. 947–951.

Ugarova, T. Yu., and Diskina, B.S., 1964. Biological activity of serum albumin of man and ox pancreatic ribonuclease in the artificially prepared complexes of these proteins with DNA. Biokhimiya, Vol. 29, No. 5, pp. 914–923.

Ugolev, A.M., 1960a. On the existence of membrane (contact) digestion. Bull. Eksperim. Biol. i Med., Vol. 49, No. 1, pp. 12–17.

Ugolev, A.M., 1960b. Certain general regularities of the cavital and membrane digestion. Works of Scientific Conference on the Problem of Physiology and Pathology of Digestion, dedicated to the memory of Academician K.M. Bykov. Ivanovo, pp. 829–834.

Ugolev, A.M., 1960c. Influence of the surface of the small intestine on enzymatic hydrolysis of starch. Nature, Vol. 188, No. 4750, pp. 588–589.

Ugolev, A.M., 1961a. Digestion and Its Adaptation Evolution. Vysshaya Shkola, Moscow.

Ugolev, A.M., 1961b. Data on membrane digestion. Communication III. Comparison of enzymatic hydrolysis of starch in the intestine in vitro. Bull. Eksperim. Biol. i Med., Vol. 52, No. 8, pp. 8–12.

Ugolev, A.M., 1963. Membrane (Contact) Digestion. Izd. AN SSSR, Moscow–Leningrad.

Ugolev, A.M., 1964. Interrelationship of digestive and absorption processes in the intestinal cell. Symposium "Physiology and Pathology of Absorption in the Gastrointestinal Tract." Odessa, pp. 7–20.

Ugolev, A.M., 1965a. Evolutionary aspects of membrane digestion. IV Scientific Conference on Evolutionary Physiology, dedicated to the memory of L.A. Orbeli, Leningrad, pp. 260–261.

Ugolev, A.M., 1965b. Membrane (contact) digestion. Physiol. Reviews, Vol. 45, No. 3, pp. 555–595.

Ugolev, A.M., and Chernousova, V.V., 1963. Data on the physiology of membrane (contact) digestion. Communication IV. On the cleavage of starch in the vessel channel of the frog. Bull. Eksperim. Biol. i Med., Vol. 55, No. 5, pp. 25–28.

Ugolev, A.M., and Marauska, M.K., 1964. Data on the physiology of membrane digestion. Comparison of hydrolysis of starch in the intestine and in vitro spectrophotometry of iodine-starch complexes. Bull. Eksperim. Biol. i Med., Vol. 47, No. 4, pp. 16–20.

Ugolev, A.M., and Salenietse, I.K., 1964. Digestive functions of the surface of the intestine of rabbits in the first weeks of life. Bull. Eksperim. Biol. i Med., Vol. 58, No. 7, pp. 15–18.

Ugolev, A.M., Iezuitova, N.N., and De Laey, P., 1964a. Localization of invertase activity in small intestinal cells. Nature, Vol. 203, No. 4947, pp. 879–880.

Ugolev, A.M., Iezuitova, N.N., Timofeeva, N.M., and Fedushina, I.N., 1964b. Location of hydrolysis of certain disaccharides and peptides in the small intestine. Nature, Vol. 202, No. 4934, pp. 807–809.

Ugolev, A.M., Iezuitova, N.N., Nadirova, T.Ya., and Timofeeva, N.M., 1966. Digestive functions of intestinal epithelium in severe irradiations. Dokl. Akad. Nauk SSSR, Vol. 166, No. 2, pp. 472–475.

Ugolev, A.M., Iezuitova, N.N., Timofeeva, N.M., Chernyakhovskaya, M.Yu., and De Laey, P., 1967. Schrift voor Gastro-Enterologic (Belgium), Vol. 10, No. 2, p. 152.

Ugolev, A.M., Iezuitova, N.N., Kashina, N.A., Sabsai, B.I., Spiridenkova, T.I., Timofeeva, N.M., Shchelovanova, I.N., Shalygina, T.V., and Sherstobitov, O.E., 1961. Newer data on the physiology of membrane digestion. Scientific conference on the problems of physiology and pathology of digestion and absorption. Abstracts of reports. Odessa, pp. 240–243.

Ukhtomskii, A.A., 1941. Outline of physiology of the nervous system. Collected Works, Vol. IV. Publications of Leningrad University, Leningrad (1954).

Vasseur, E., 1951. Demonstration of a jelly-splitting enzyme at the surface of the sea-urchin spermatozoan. Exptl. Cell Res., Vol. 2, pp. 144–146.

Verzär, F., and Sailer, E., 1952. Glukose-Resorption und Alkalische Phosphatase des Dünndarmes nach Adrenalektomie, bei mit NaCl Behandelten Tieren. Helv. Physiol. Pharmacol. Acta, Vol. 10, Part 2, pp. 247–258.

Warburton, F.E., 1961. Timing of stages of the mitotic cycle in the small intestine of the rat. Anat. Record, Vol. 139, No. 2, p. 284.

Weijers, H.A., van de Kamer, J.H., Mossel, D.A.A., and Dicke, W.K., 1960. Diarrhea, caused by deficiency of sugar-splitting enzymes. Lancet, Vol. II, No. 7145, pp. 296–297.

Wheeler, E., 1955. The rate of reactions and the selectivity in the pores of the catalyst. In: Catalysis. Theory and Methods of Investigation. IIL, Moscow, pp. 479–563.

Widner, W.R., Storer, J.B., and Lushbaugh, C.C., 1951. The use of x-ray and nitrogen mustard to determine the mitotic and intermitotic time in normal and malignant rat tissues. Cancer Res., Vol. 11, pp. 877–884.

Wiggans, D.S., and Johnston, J.M., 1959. The absorption of peptides. Biochim. et Biophys. Acta, Vol. 32, No. 1, pp. 69–73.

Wilkes, B.G., and Palmer, E.T., 1932. Similarity of the kinetics of invertase action in vivo and in vitro, II. J. Gen. Physiol., Vol. 16, No. 2, pp. 233–242.

Williams, A.W., 1963. Electron microscopic changes associated with water absorption in the jejunum. Gut, Vol. 4, No. 1, pp. 1–7.

Willstätter, R., and Lowry, C.A., 1925. Invertinvermiderung in der Hefe. Elfte Abhandlung zur Kenntnis des Invertins. Hoppe Seyler's Z.Physiol. Chem., Vol. 150, No. 2, pp. 287–305.

Wilson, T.H., 1962. Intestinal Absorption. Philadelphia–London.

Wilson, T.H., 1965. Absorption of carbohydrates and protein. In: Current Concepts of Clinical Gastroenterology, pp. 93–105.

Wilson, T.H., and Wilson, D.W., 1958. Studies in vitro of digestion and absorption of pyrimidine nucleotides by the intestine. J. Biol. Chem., Vol. 233, No. 6, pp. 1544–1547.

Wilson, T.H., and Wiseman, G., 1954. The use of sacs everted small intestine for the study of the transference of substances from the mucosal to the serosal surface. J. Physiol., Vol. 123, No. 1, pp. 116–125.

Wilson, T.H., and Vincent, T.N., 1955. Absorption of sugars in vitro by the intestine of the golden hamster. J. Biol. Chem., Vol. 216, No. 2, pp. 851–866.

Windle, W.F., 1940. Physiology of the Fetus. Philadelphia–London.

Yonge, C.M., 1937. Evolution and adaptation in the digestive system of the Metazoa. Biol. Reviews, Vol. 12, No. 1, p. 87.

Yuile, C.L., O'Dea, A.E., Lucas, F.V., and Whipple, G.H., 1952. Plasma protein labeled with lysine-ε-C^{14}. Its oral feeding and related protein metabolism in the dog. J. Exptl. Med., Vol. 96, No. 3, pp. 247–254.

Zetterqvist, H., 1956. The Ultrastructural Organization of the Columnar Absorbing Cells of the Mouse Jejunum. Stockholm.

Zetterqvist, H., and Hendrix, T.R., 1960. A preliminary note on an ultrastructural abnormality of the intestinal epithelium in adult celiac disease (nontropical sprue) which is reversed by a gluten-free diet. Bull. Johns Hopkins Hosp., Vol. 106, No. 3, pp. 240–249.

Intestinal Biopsy. New York (1962).

Manual in Cytology, Vol. 1. Nauka, Moscow–Leningrad (1965).

Symposium "Origin of Prebiological System." Mir, Moscow (1966).

GPSR Compliance

The European Union's (EU) General Product Safety Regulation (GPSR) is a set of rules that requires consumer products to be safe and our obligations to ensure this.

If you have any concerns about our products, you can contact us on ProductSafety@springernature.com

In case Publisher is established outside the EU, the EU authorized representative is:

Springer Nature Customer Service Center GmbH
Europaplatz 3
69115 Heidelberg, Germany

Batch number: 10704707

Printed by Printforce, the Netherlands